MW00462847

REFLECTIONS ON THE MOTIVE POWER OF FIRE

Sadi Carnot at the age of 34.

REFLECTIONS ON THE MOTIVE POWER OF FIRE

SADI CARNOT

and other Papers on the Second Law of Thermodynamics by
É. Clapeyron and R. Clausius

Edited with an Introduction by
E. Mendoza

DOVER PUBLICATIONS, INC.
Mineola, New York

Bibliographical Note

Reflections on the Motive Power of Fire is an unabridged and slightly corrected republication of *Reflections on the Motive Power of Heat* by Sadi Carnot, translated and edited by R. H. Thurston, and published by Macmillan and Company in 1890. Additional footnotes by E. Mendoza are initialled "E. M." A new Appendix, "Selections from the Posthumous Manuscripts of Carnot," translated by R. H. Thurston and E. Mendoza, has been added.

Memoir on the Motive Power of Heat by É. Clapeyron was especially translated for this edition by E. Mendoza.

On the Motive Power of Heat, and on the Laws which can be Deduced from it for the Theory of Heat by R. Clausius was translated by W. F. Magie, and originally appeared in the volume entitled *The Second Law of Thermodynamics* edited by W. F. Magie, and published by Harper and Brothers in 1899.

E. Mendoza edited and wrote the Introduction to this Dover edition.

This is a new edition of the work first published in this form by Dover Publications, Inc. in 1960.

International Standard Book Number: 0-486-44641-7

Manufactured in the United States of America
Dover Publications, Inc., 31 East 2nd Street, Mineola, N.Y. 11501

Contents

Illustrations

Introduction

§ 1. *Sadi Carnot.* The beginning of the nineteenth century was a time which resembled our own in many ways. Men were aware that it was a period of transition and that they were entering a new era—the age of steam power and the age of iron. The first, and at the time the most important use of the new motive power was for driving pumps for draining coal mines; coal was desperately needed by both England and France for making the coke required for smelting iron. A secondary use was for driving machinery (the bellows and rolling mills of the iron works, the paddles of steam-boats); a use which the recent invention of various cranking mechanisms had made possible. In exploiting these applications, England, relying on the genius of engineers who had no formal training whatever, was well ahead of France. The French realized that England's prosperity in peacetime and her invulnerability in war were both due to her industrialization. In France there was a great awareness of the growing importance of science, and there, for perhaps the first time in history, a real attempt was made to give engineers and scientists a specialized formal training, while the more enlightened of her governments sponsored quite elaborate researches.

One of the most important of the institutions established in the course of this movement was the École Polytechnique, founded in 1794 as a training school for army engineers. A list of men associated with its early period, before 1830, reads like the index of a book on mathematical physics. Among the first instructors were Lagrange, Fourier, Laplace, and Berthollet, Ampère, Malus and Dulong; among former students who stayed on as instructors were Cauchy, Arago, Désormes, Coriolis, Poisson, Gay-Lussac, Petit and Lamé; other students included Fresnel, Biot, Sadi Carnot, Clapeyron and the physiologist Poiseuille. It was this generation which largely formulated the attitudes and procedures of mathematical physics as we know it today. The terms Lagrangian, Laplace transform, Poisson's equation, Fourier integrals, Cauchy relations, Fresnel coefficients, Coriolis forces conjure up a picture of an era when it seemed that the understanding of nature was merely a question of writing down equations and of inventing methods for solving them.

But there was another aspect of the work of this extraordinary group of scientists. Possibly because of their connection with an army which conquered most of Europe, possibly because they lived in such a time of discovery and transition, almost all of them saw their science as only one part of their lives. While being masters of pure science, they applied their astonishingly varied talents to very different problems. For example, students of history think of Fourier not as a mathematician but as one of Napoleon's generals in Egypt. Again, Clément and Désormes published three papers in the same volume of the *Journal de Physique* for 1819: one described their work on the specific heats of gases and the absolute zero of temperature; the second was an assessment of the food requirements of the country with proposals for the storage of dehydrated grain in wrought-iron granaries; and the third dealt with the relative merits of coal-gas and oil for domestic lighting. Clapeyron and Lamé built some of the first railroads in France, and were responsible for much of the organization of technical education in Russia.

One of the outstanding men of this generation, one of the great men of history, was Lazare Carnot. He was the man who appointed Napoleon to his first independent command, who organized the fourteen armies which for a time conquered Europe, who planned every campaign and who, at Antwerp, remained the only one of Napoleon's generals to be undefeated. In addition, he was an excellent mathematician. He wrote on abstract geometry and on the logical problems of the calculus, but from our present point of view it is more interesting to recall a mathematical work entitled *Fundamental Principles of Equilibrium and Movement*, published in 1803. It is a discussion of the efficiency of machines—machines in the old sense, that is—such as pulleys and inclined planes. His contribution was to see through all the details of the mechanisms and to develop a much more general discussion based on what we should now call the conservation of mechanical energy. It seems trivial enough to us now, but in those days it must have been a very bold approach. As late as the 1830's, the term "Carnot's theorem" denoted a statement that in any machine the accelerations and shocks of the moving parts all represented losses of "moment of activity," that is, of the useful work done; from this Carnot drew the inference that perpetual motion was impossible.

In 1796, when Lazare Carnot was one of the Directory of France, a son was born. He was named Sadi after a mediaeval Persian poet and moralist, Saadi Musharif ed Din, whose poems had enjoyed something of a vogue. A few years later Lazare was Minister of War and he often took his small son with him when he visited Napoleon. It is recorded that one afternoon Napoleon's wife and some other ladies were in a boat on a lake and Napoleon was amusing himself throwing stones near them and splashing them; they dared not protest. Sadi, all of four years old, ran up and shook his fist, shouting: "You beastly First Consul, stop teasing those ladies!" Luckily for the history of physics, Napoleon roared with laughter.

Till he was about 16, Sadi's education was directed by his father; but after that time they were able to see one another only on the briefest occasions. Sadi entered the École Polytechnique in 1812—the year when Napoleon's fortunes turned, the year of his retreat from Moscow. By the time Sadi was finishing his courses of study, in 1814, Paris itself was being besieged. The students of the École Polytechnique petitioned to be allowed to take part in the defence of the city; 250 of them were sent to a skirmish against the Prussians at Vincennes. They had 28 cannon and fought bravely enough, repelling several charges, but what looked like a relief force turned out to be a regiment of Russian lancers and the contingent retired in confusion, though with no losses.

A few weeks later Napoleon abdicated and a violently anti-republican king was put on the throne. Later in the year Sadi left the École Polytechnique to join the Engineers. For much of his life he remained an army officer, bearer of a family name alternately illustrious and ominous, according to the bewildering changes of the French political scene. For a while, during the Hundred Days when Napoleon escaped and Lazare Carnot was on the scene again, Sadi was disgusted to find that his humble junior-lieutenant's room was visited by high-ranking officers anxious to curry favor. But when the monarchy was restored and Lazare exiled, Sadi found himself on garrison duty, far from Paris, doing the lowest routine jobs. Later he transferred to the General Staff, but almost immediately retired on half-pay and moved to Paris; this was in 1820 and he was 24 years of age.

The period that followed was the creative time of his life. He studied widely—at the Sorbonne, the Collège de France, the École des Mines and elsewhere—concentrating on physics and economics.

He spent much of his time visiting factories and studying the organization and economics of various industries; he became an expert on the industries and trades of different countries of Europe. Once he managed to visit his father in Magdeburg for a few weeks. On that occasion Lazare said to him; "If real mathematicians were to take up economics and apply experimental methods, a new science would be created—a science which would only need to be animated by the love of humanity in order to transform government."

He was reserved, almost taciturn, with a hatred of any form of publicity. Among his notes were found pages of rules of conduct which he jotted down; a few of these may give an idea of his character. "Say little about what you know and nothing at all about what you don't know. . . . Why not say more often 'I don't know?' . . . When a discussion degenerates into a dispute, Keep silent. . . . Do not do anything which the whole world cannot know about. . . ." Apparently he managed to live up to these high standards and his friends all spoke of his underlying warmth and humanity. Passionately fond of music, he was an excellent violinist who preferred the classical Lully to the "moderns" of the time; he was devoted to literature and all the arts.

In 1823, Lazare died in exile and Hippolyte, his younger son, returned to Paris. The two brothers set up home in a small apartment and its was here that Sadi wrote the *Reflections on the Motive Power of Fire*. He made Hippolyte read parts of the manuscript to satisfy himself that they would be intelligible to the non-scientist. There is little doubt that it was intended as a popular book, not a technical treatise. The text contains no arguments which depend on a calculus treatment—those are confined to the footnotes—but mostly verbal statement couched in simple but exact language. The main theme is the design of engines, with a good deal of emphasis on their importance to the French nation. High- and low-pressure steam-engines, air-engines, and an internal-combustion machine are all examined critically. But the Memoir transcended technical details because Sadi had inherited from his father the capacity to generalize, to see the fundamental processes animating a complicated mechanism. Thus he saw that in an engine—any engine—an amount of caloric fell from a high to a low temperature; he extended some of his father's ideas on mechanics to apply to thermal processes—the impossibility of perpetual motion, the need to avoid irreversible changes.

Above all, he invented the closed cycle of operations. The

importance of this advance can best be gauged by comparing Carnot's well-known discussion with Petit's comparison of the efficiencies of air- and steam-engines, published in 1818. Petit began by proving that the work done in an isothermal expansion between volumes V_1 and V_2 was $RT \log V_2/V_1$. This formula he applied to 1 gram of water turning to steam, the volume ratio and the latent heat being known; then, assuming 1 gram of air to absorb the same quantity of heat, its temperature would rise by an amount determined by its specific heat at constant pressure and this was equivalent to a certain volume expansion. Hence it emerged that an air-engine was enormously more efficient than a steam-engine. It is difficult to make a complete list of all the *non sequiturs* in the argument, yet it was the best that French analytical physics had yet produced.

Sadi's Memoir was published as a small book in 1824 and sold for 3 francs. It received one long and excellent review, in the *Revue Encyclopédique*, a journal which covered all branches of literature. It ended: "Monsieur Carnot is not afraid of tackling difficult questions; and in this first production he shows himself capable of going into a matter which has become today one of the most important with which theoreticians and physicists can occupy themselves." But apparently hardly anyone bought the book; a few years later booksellers had never even heard of it.

In the same year, 1824, the political scene worsened with the accession of a king who did his best to restore absolute monarchy. For a time, Sadi was recalled to full-time service, as a staff captain. But in 1828 he resigned permanently, and devoted himself to physics and economics. His physics notebooks of this period have been preserved and they make fascinating reading; while many others had speculated on the equivalence of energy and heat, his penetrating mind could see the logical consequences. One fragment of his economic writing has also been found; he envisaged the use of taxes not merely as a source of revenue, but also as a means of guiding the agricultural and industrial development of the country. In 1830 there was a revolution and a more progressive king was installed on the throne. His friends thought that Sadi would be called to a high position in the government; but there were intrigues against him because of his republican record and he was passed over. Sadi became convinced that only by some fundamental change of heart could political practice advance; he interested himself in the Association Polytechnique, a society, formed by former students of the École, which aimed at popular scientific education.

In August, 1831, he started some kind of experimental research, probably vapor-pressure measurements on steam. At the beginning of June of the next year he was in Paris while an anti-government riot was going on, in which the École Polytechnique students were taking a noisy part. He saw a drunken officer galloping down the street brandishing his sabre and knocking people down; Sadı dodged under the man's arm, toppled him off his horse and threw him in the gutter. Two weeks later Sadi caught scarlet fever which turned into a brain fever; he recovered and was taken into the country to convalesce. Hippolyte and another friend went to nurse him. He made a melancholy study of the reports of the cholera epidemic then raging. Some days later he himself caught the disease and died in a few hours. He was just 36 years old.

Carnot's *Reflections* had been received by the official world of science in utter silence. No contemporary reference to his work can be traced in any published paper; his personal friends seem to have been the only people who had read his book and who now mourned the passing of a great physicist. By a bizarre coincidence, another brilliant young innovator died in Paris only a few days before Carnot, equally unrecognized and rejected by those in authority—Evariste Galois, probably the most original mathematician in France at the time; it was Fourier who had lost one of the papers he submitted to the Academy, Poisson who had rejected the other. One can hardly avoid the conclusion that the men who had in their youth transformed physics and mathematics, had degenerated into venerable Academicians, no longer receptive to new ideas coming from younger men.

The only person who kept the memory of Carnot's results alive was Émile Clapeyron. In the year of their publication, he had been in Russia with Lamé; he returned to Paris in 1830. Clapeyron had written extensively on elasticity and the design of bridges, on the organization of public projects and on technical education, before he published his "Memoir on the Motive Power of Heat" in the *Journal de l'École Polytechnique* in 1834.

Clapeyron's paper was quite different from Carnot's in its manner of presentation, though it reached the same results. It was analytical throughout and made only cursory references to the problems of engine design and to the industrial applications which had been prominent in the original; and where Carnot had been simple and lucid, he tended to be repetitive. But in this paper Clapeyron also deduced new results and made their derivation clear through the use of indicator diagrams. Practically all his equations are correct whereas many of Carnot's are incorrect however they are construed. Like its predecessor, this Memoir attracted little attention; one of the few definitely incorrect formulae in it was misapplied to interpret some incorrect measurements of the specific heat of air by von Suerman. But in 1843 Clapeyron's paper was translated into German and republished in the *Annalen der Physik und Chemie.* It was largely through this translation that Carnot's results first reached physicists.

Carnot and Clapeyron wrote their papers at a time when there was no unambiguous statement of the equivalence of heat and energy. As a result, many of their statements seem to be incorrect at first reading. However, the difficulty is largely one of understanding the different ways in which they used the words "heat" and "caloric"; the rest of this Introduction will be devoted to this problem.

§ 2. *Fire, Heat and Caloric.* As late as 1783, Montgolfier could state that his hot-air balloon ascended because it was filled with fire, the lightest of the four elements. This ancient theory was finally abandoned in scientific circles after Lavoisier's work on combustion, though the picturesque use of the word "fire"—for example, in Carnot's title—continued long after.

The terms "heat" and "caloric" must be discussed in more detail. In the English-speaking world, the account usually given of the caloric theory of heat derives directly from that given in his later years by Kelvin. It is well known that in this theory, heat phenomena were visualized as due to a gas called caloric, whose presence in the pores between the atoms of a solid caused thermal expansion and whose emission from surfaces explained Newton's law of cooling. Rumford's work was taken to mean that this gas was almost weightless, and that it could be squeezed out "like water from a sponge" when work was done on a body. However, since Kelvin himself had changed from a supporter of the theory to an opponent of it, he tended to emphasize the contrasts between this approach and the modern one, and to pour scorn on the caloric theory. Thus we have inherited from him a distorted perspective.

The great majority of British "modern" physicists were convinced of the correctness of the caloric theory. Their attitude is summarized in Nicholson's *Journal* for 1807: ". . . it is well known that Count Rumford adheres to the old theory of heat being simply a vibratory motion of the particles of bodies." Even Davy subscribed to the caloric theory, possibly because he knew his famous ice-rubbing experiment was unsatisfactory, possibly because all the chemists regarded caloric as an essential constituent of oxygen. But in France the situation was different.

It is convenient to limit the discussion to the period between 1780, when the *Mémoire sur la Chaleur* was published by Lavoisier and Laplace, and 1836, when Lamé's *Cours dé Physique* was published as a textbook for use in the École Polytechnique. There are several textbooks, many teaching articles in the *Journal de l'École Polytechnique*, and very many research papers between these dates, from which can be formed a good idea of the theories which were current while Carnot and Clapeyron were students and later were formulating their own theories.

The most important fact which emerges is that the caloric theory, which implied that heat was conserved in all thermal processes, and the theory that heat was equivalent to work were *both regarded as true* by French physicists. All textbooks and every teaching article invariably presented the two theories side by side; they were generally believed to be different aspects of the same explanation. To take but two examples spanning the chosen period; Laplace and Lavoisier stated in 1780: "In general, one can change the first hypothesis into the second by changing the words 'free heat, combined heat, and heat released' into '*vis viva*, loss of *vis viva*, and increase of *vis viva*.'" In 1836, Lamé wrote in his book: ". . . we will continue to understand by quantity of heat, the energy or the intensity of the unknown cause of the changes of density and of state of ponderable bodies. In the emission hypothesis, this quantity is the mass of the caloric; in that of undulations, it is the *vis viva* of the movements propagated, or the square of the amplitude of the vibrations." It is true that for the purposes of calculation the caloric theory was always used, but that was merely because it lent itself more easily to calculation.

In fact, there are *two* heat quantities that it is useful to define. One is conserved in any reversible process, whether work is done or not—in modern terms, the entropy; the other is conserved by a body in adiabatic calorimetry—in modern terms, the quantity of heat or heat energy. Neither quantity is any more fundamental than the other. It is through an historical accident, an arbitrary choice, that we happen to call one of these quantities by the familiar term "heat" and the other by a pseudo-Greek name. In the early nineteenth century the two were confused. The writers of that time also had two names for the heat available, but when we read *any* paper of this period, "quantity of heat" and "quantity of caloric" must each be construed to mean sometimes "change of entropy" and sometimes "quantity of heat." Carnot himself stated explicitly that he used the two terms interchangeably; in fact, he had a tendency to reserve the term "caloric" for situations where we should now talk of entropy. For example, he usually wrote that in a reversible heat-engine, the amounts of caloric absorbed by the working substance at the high temperature and lost at the low were compensated. Clapeyron, for the most part, adopted the other usage, calling this "quantity of heat." But neither writer—nor any other of the same period—was entirely consistent; nor is this surprising in view of the poor experimental data then available. In practice it is difficult to distinguish between conservation of entropy and conservation of energy in experiments over a small temperature range. It even happened that the standard experimental procedure for determining what we now call quantities of heat, using the ice calorimeter, could equally well be said to have measured changes of entropy since the experiment took place at constant temperature.

But if the terminology was confused, the equations which could be written down were quite definite. It is interesting, therefore, to go a little more deeply into the quantitative aspects of the caloric theory and the theoretical models which were proposed for it, and at the same time into the difficulties of visualizing and analyzing the equivalence of heat (or caloric) and work.

The total mass of caloric in a body—called the absolute heat—was regarded as a mixture of two components; the free or perceptible caloric (*calorique sensible*) could affect a thermometer while the other component, the latent caloric, could not. (There is in many ways a strong resemblance to the modern two-fluid description of superfluid liquid helium.) The use of the words "latent caloric" implied a similarity to latent heat in the more usual sense; it was regarded as chemically bound to the molecules of the body. In the adiabatic compression of a gas, the absolute heat remained constant but the observed rise of temperature indicated that some latent caloric had become perceptible. Equations purporting to express this analytically were given by Poisson in 1833 (following Laplace in his *Méchanique Céleste* of 1823). He argued that the state of a given mass of gas was uniquely determined by two co-ordinates—for example, the pressure p and volume v. Then the change of absolute heat dq was a perfect differential and could be written

$$dq = \left(\frac{\partial q}{\partial p}\right)_v dp + \left(\frac{\partial q}{\partial v}\right)_p dv$$

in modern notation. (Partial derivatives were then written like ordinary ones.) Then

$$\left(\frac{\partial q}{\partial p}\right)_v = \left(\frac{\partial q}{\partial \tau}\right)_v \cdot \left(\frac{\partial \tau}{\partial p}\right)_v = \frac{V \cdot C_v}{R}$$

for a perfect gas, where C_v is the specific heat at constant volume, and similarly for the other term. For adiabatic expansion, $dq=0$; the correct, familiar expression $pv^\gamma=$constant emerges for the process. The great triumph of this calculation was the prediction of the correct velocity of sound—the prediction that the ratio of the adiabatic and isothermal elasticities was the ratio of the specific heats—using the measurements of γ of Clément and Désormes and of Gay-Lussac and Welter. But it will be seen at once that the two-fluid model is quite irrelevant to the calculation.

Unfortunately, it was possible to make another deduction from the equations, namely that the specific heat of a perfect gas decreased when it was compressed. Poisson gave the explicit formula $p^{1/\gamma-1}$ for this variation. This appeared quite plausible on the two-fluid model and it gave an alternative interpretation of the temperature changes in adiabatic processes. It was particularly unfortunate that Delaroche and Bérard (whose very fine specific heat measurements under constant pressure were standard till Regnault's work forty years later) had found this result experimentally. A given volume of air at 1006 mm. pressure was found to have a heat capacity 1.240 times that at 740 mm. (the mean of two readings), though the pressure ratio was 1.358. The effect was, of course, spurious and it was difficult to reconcile it with the apparent constancy of specific heats with temperature; probably no other bad data have upset the development of thermodynamics more than these. Both Carnot and Clapeyron were led astray by them; both interpreted their expressions for the absolute heat of a perfect gas to mean that the specific heat varied as the logarithm of the volume. Had Delaroche and Bérard's measurement at a pressure above atmospheric been more accurate, they would have been led to very different conclusions.

Remembering the need to interpret the terms heat and caloric properly, the caloric theory is seen to have been an excellent basis for calculation. In most of Carnot's formulae C_p and C_v should be replaced by C_p/T and C_v/T; thus his equations for isothermal processes are correct in form, but those for adiabatic processes need modification. Clapeyron's equations need little modification since he incorporated an unknown function of temperature $C(T)$ which is, in fact, proportional to the absolute temperature.

One *model* of the caloric gas was widely accepted. It was regarded as radiant heat—which was, of course, composed of particles. Thus a body at a certain temperature was replaced by a constant temperature enclosure inside which the radiation-gas was in dynamic equilibrium. When the particles of radiation were combined with the molecules of the body, the caloric was latent; when in transit between molecules it was perceptible: the density of the particles of radiation in transit determined the temperature of the body. This picture had been developed largely by Fourier to account for heat conduction. Laplace took it very seriously and used it to calculate expressions for the absorption and reradiation coefficients of the molecules. Later the development of the wave theory of light by Fresnel and others undermined some of the confidence in the caloric theory of heat.

It is instructive to consider also a variant view—that held by Clément and Désormes (as well as Lambert and John Dalton). They argued—rather tenuously, it seems to us—that since heat processes could take place in a vacuum, then vacuum must have its own heat capacity per unit volume. Now a volume of air at low pressure could be considered to be composed of a smaller volume of air at standard pressure, mixed with a certain volume of vacuum. When air was allowed to rush into the space, vacuum was destroyed, its total caloric was given up and the temperature rose; and the equation of an adiabatic compression deduced on this model is of the correct analytical form. Clément and Désormes expressed their specific heat results as the ratio of the specific heat per unit volume of vacuum to that of air at 1 atmosphere pressure; it was left to Laplace to recast their results in terms of the ratio γ. On the assumption that the specific heat of vacuum was constant with temperature, they located the position of absolute zero at about 250°C, by using different expansion ratios. All Clément and Désormes had done was to substitute the words "caloric of vacuum" for the "latent caloric" used by the majority of physicists. Nevertheless, their Memoir was rejected and they became very bitter about the reception of their work.

The equations governing an adiabatic expansion are, in fact, independent of the model which is used for the process. They depend only on the existence of an equation of state and the definition of two suitable specific heats, and, as we have seen, the correct relations were deduced on two incorrect models. It is something of a puzzle to know why Carnot and Clapeyron, who both possessed great analytical ability, did not accept these equations. Clapeyron, for example, specifically stated that Laplace and Poisson used hypotheses which could be disputed, and that the law relating volume and temperature was unknown. Perhaps he had no confidence in the equation because the same set of assumptions did not allow the p–T relations to be calculated for a saturated vapor like steam.

Having considered the caloric theory, it is interesting now to try to understand some of the difficulties of developing the alternative theory that heat was the *vis viva* of the molecules of a body. One important obstacle was that at the time the accepted picture of the molecular motions in a gas was quite incorrect. As early as 1738 Bernoulli had proposed a correct qualitative explanation of the pressure exerted by a gas in terms of the impulses of the fast-moving molecules, but the French physicists ignored this kinetic model. Instead, they considered a gas to be a sort of highly rarified solid, the pressure on the walls being a static effect of the intermolecular repulsions; the only movements of the molecules—apart from convection currents—were small oscillations about their mean positions. Solids, where convection was entirely absent, therefore seemed to be simpler systems to deal with, but it is not surprising that they found the analysis of the motions intractable. Though the French mathematicians had shown that almost every phenomenon in nature could be reduced to a problem in central forces, here they could hardly begin. Poisson went so far as to call the theory sterile. Even had they accepted the macroscopic equivalence of work and heat, they would have had to concentrate solely on the energy aspect of the phenomena, and forget the entropy aspect. Nor was there any guarantee that the conversion factor connecting heat and work was constant with temperature. Indeed Kelvin himself, in an early paper discussing the Clapeyron equation for the latent heat of steam, deduced that J actually did vary with temperature. Further, in order to preserve the identity with the caloric picture, it was assumed that the velocity of propagation of the atomic vibrations was that of light. So this model was no more economical of hypotheses than the caloric theory: one postulated particles of radiant heat, the other needed an ether to propagate the undulations. In his private notes, Carnot has left us a detailed account of some of his difficulties with the theory. He accepted easily the macroscopic equivalence of heat and work and proposed many experiments—including the Joule–Kelvin experiment—to test it critically. But he could not visualize what form the atomic vibrations could take in a solid; like most scientists he probably imagined the atoms to be stacked together touching one another, and his best suggestion for the vibrations was that the atoms "changed places." But there were other difficulties. If heat were *vis viva*, then *vis viva* was obviously conserved on the universal scale; therefore, there could be no friction between atoms, so that atoms could not touch one another. But if this were so, it was impossible for him to visualize the forces which held them apart. It was no use postulating that the ether held the atoms in position because the ether was a fluid and therefore itself atomic in structure, and the difficulty was only postponed. A final serious difficulty was that if heat were the vibration of atoms, why was a cold body necessary to extract work from an engine? The answer had been obvious on the theory which could use entropy as its heat quantity, but Carnot never clearly distinguished between the two functions. Finally he did accept the *vis viva* theory and made an estimate of J, but he never did any work with it.

It was through the speculations of a number of German scientists —Mohr, Mayer, Holtzmann and Clausius—and the electrical and other experiments of Joule that this view was adopted. It was Joule also who suggested to Kelvin that Clapeyron's $C(T)$ was proportional to the absolute temperature. The name "heat" was firmly attached to the energy content (though Kelvin suggested years later that it be renamed "caloric" lest there be any confusion with popular ideas) and when the other heat function was reintroduced by Clausius shortly afterwards, it was so clearly differentiated from heat that it appeared to be something entirely new.

After 1850 a number of papers were published which rewrote Carnot's results in the new terminology. Of these, Clausius' has been selected for this volume—its terseness is more attractive than Kelvin's rather pompous style. (It contrasts even more strongly with other German writings of about the same period; the famous paper by Mayer on the equivalence of work and heat, for example, is almost unreadable today, the physical ideas being obscured under layers of philosophical verbiage.) With this paper by Clausius and the one two years later in which he defined entropy, thermodynamics took the form we know today.

§ 3. *Bibliography and Notes on the Translations.* Sadi Carnot's obituary notice was written by his "most intimate friend," Robelin, and published in the *Revue Encyclopédique* LV, 528 (1834). Most of the details of his life come from the writings of his brother, Hippolyte; a few are given in his biography of Lazare, *Mémoires sur Carnot* (1861–63). In 1868, the Conte di St. Robert wrote to the Carnot family to find out biographical details. Hippolyte's son, also called Sadi, sent a long letter which St. Robert published in the *Atti della R. Accademia delle Scienze di Torino* IV, 151 (1868). The *Comptes Rendus* of the Paris Académie des Sciences LXVIII, 115 (1869) contains an account by Sadi's fellow-student, M. Chasles, of the battle at Vincennes. In 1878 the *Reflections* was republished with a biography by Hippolyte, largely a rewrite of the 1868 article with some additional anecdotes and extracts from Sadi's private writings.

The best-known edition of the *Reflections* is that published in 1878 by Gauthier-Villars; the original, old-fashioned spelling was modernized in that edition. The original text has been reproduced photographically more than once, the latest publication being by Blanchard (Paris) 1953. The physics notebooks which were found posthumously have been reproduced photographically and published under the title *Sadi Carnot, biographie et manuscrit* by Gauthier-Villars (Paris) 1927, with a transcript and brief commentary and Hippolyte's biography.

Commentaries on the use of the words "caloric" and "heat" have been given by H. L. Callendar, *Proc. Phys. Soc.* (London) XXIII, 153 (1911) and V. K. La Mer, *Am. J. Phys.* XXII (1954) and XXII, 95 (1955), etc. I should also like to acknowledge my indebtedness to Prof. F. C. Frank of Bristol University and Prof. L. Rosenfeld of Manchester University for interesting conversations.

In the translations of the Carnot and Clapeyron memoirs the words "caloric" and "heat" are carefully used as in the original French, except in the phrase *calorique spécifique* (occasionally used as an exact alternative to *chaleur spécifique*) which is translated as "specific heat." *Machine à feu* is translated as "heat-engine," *force vive* as "*vis viva.*" Carnot's misspellings of British surnames—Smeathon, Trevetick, Robinson for Smeaton, Trevithick and Robison—have been corrected.

E. MENDOZA

Manchester University,
March, 1960.

REFLECTIONS ON THE MOTIVE POWER OF FIRE AND ON MACHINES FITTED TO DEVELOP THAT POWER

BY SADI CARNOT

one-time pupil of the

ÉCOLE POLYTECHNIQUE

1824

Translated and edited by

R. H. THURSTON

RÉFLEXIONS

SUR LA

PUISSANCE MOTRICE

DU FEU

ET

SUR LES MACHINES

PROPRES A DÉVELOPPER CETTE PUISSANCE.

PAR S. CARNOT,

ANCIEN ÉLÈVE DE L'ÉCOLE POLYTECHNIQUE.

A PARIS,

CHEZ BACHELIER, LIBRAIRE,

QUAI DES AUGUSTINS, N°. 55.

1824.

Title page of the memoir published in 1824.

Reflections on the Motive Power of Fire, and on Machines Fitted to Develop that Power

Every one knows that heat can produce motion. That it possesses vast motive-power no one can doubt, in these days when the steam-engine is everywhere so well known.

To heat also are due the vast movements which take place on the earth. It causes the agitations of the atmosphere, the ascension of clouds, the fall of rain and of meteors, the currents of water which channel the surface of the globe, and of which man has thus far employed but a small portion. Even earthquakes and volcanic eruptions are the result of heat.

From this immense reservoir we may draw the moving force necessary for our purposes. Nature, in providing us with combustibles on all sides, has given us the power to produce, at all times and in all places, heat and the impelling power which is the result of it. To develop this power, to appropriate it to our uses, is the object of heat-engines.

The study of these engines is of the greatest interest, their importance is enormous, their use is continually increasing, and they seem destined to produce a great revolution in the civilized world.

Already the steam-engine works our mines, impels our ships, excavates our ports and our rivers, forges iron, fashions wood, grinds grains, spins and weaves our cloths, transports the heaviest burdens, etc. It appears that it must some day serve as a universal motor, and be substituted for animal power, waterfalls, and air currents.

Over the first of these motors it has the advantage of economy, over the two others the inestimable advantage that it can be used at all times and places without interruption.

If, some day, the steam-engine shall be so perfected that it can be set up and supplied with fuel at small cost, it will combine all desirable qualities, and will afford to the industrial arts a range the extent of which can scarcely be predicted. It is not merely that a powerful and convenient motor that can be procured and carried anywhere is substituted for the motors already in use, but that it causes rapid extension in the arts in which it is applied, and can even create entirely new arts.

The most signal service that the steam-engine has rendered to England is undoubtedly the revival of the working of the coal mines, which had declined, and threatened to cease entirely, in consequence of the continually increasing difficulty of drainage, and of raising the coal.* We should rank second the benefit to iron manufacture, both by the abundant supply of coal substituted for wood just when the latter had begun to grow scarce, and by the powerful machines of all kinds, the use of which the introduction of the steam-engine has permitted or facilitated.

Iron and heat are, as we know, the supporters, the bases, of the mechanic arts. It is doubtful if there be in England a single industrial establishment of which the existence does not depend on the use of these agents, and which does not freely employ them. To take away today from England her steam-engines would be to take away at the same time her coal and iron. It would be to dry up all her sources of wealth, to ruin all on which her prosperity depends, in short, to annihilate that colossal power. The destruction of her navy, which she considers her strongest defence, would perhaps be less fatal.

The safe and rapid navigation by steamships may be regarded as an entirely new art due to the steam-engine. Already this art has permitted the establishment of prompt and regular communications across the arms of the sea, and on the great rivers of the old and new continents. It has made it possible to traverse savage regions where before we could scarcely penetrate. It has enabled us to carry the fruits of civilization over portions of the globe where they would else have been wanting for years. Steam navigation brings nearer together the most distant nations. It tends to unite the nations of the earth as inhabitants of one country. In fact, to lessen the time,

* It may be said that coal-mining has increased tenfold in England since the invention of the steam-engine. It is almost equally true in regard to the mining of copper, tin, and iron. The results produced in a half-century by the steam-engine in the mines of England are today paralleled in the gold and silver mines of the New World—mines of which the working declined from day to day, principally on account of the insufficiency of the motors employed in the draining and the extraction of the minerals.

the fatigues, the uncertainties, and the dangers of travel—is not this the same as greatly to shorten distances?*

The discovery of the steam-engine owed its birth, like most human inventions, to rude attempts which have been attributed to different persons, while the real author is not certainly known. It is, however, less in the first attempts that the principal discovery consists, than in the successive improvements which have brought steam-engines to the conditions in which we find them today. There is almost as great a distance between the first apparatus in which the expansive force of steam was displayed and the existing machine, as between the first raft that man ever made and the modern vessel.

If the honor of a discovery belongs to the nation in which it has acquired its growth and all its developments, this honor cannot be here refused to England. Savery, Newcomen, Smeaton, the famous Watt, Woolf, Trevithick, and some other English engineers, are the veritable creators of the steam-engine. It has acquired at their hands all its successive degrees of improvement. Finally, it is natural that an invention should have its birth and especially be developed, be perfected, in that place where its want is most strongly felt.

Notwithstanding the work of all kinds done by steam-engines, notwithstanding the satisfactory condition to which they have been brought today, their theory is very little understood, and the attempts to improve them are still directed almost by chance.

The question has often been raised whether the motive power of heat† is unbounded, whether the possible improvements in steam-engines have an assignable limit—a limit which the nature of things will not allow to be passed by any means whatever; or whether, on the contrary, these improvements may be carried on indefinitely. We have long sought, and are seeking today, to ascertain whether there are in existence agents preferable to the vapor of water for developing the motive power of heat; whether atmospheric air, for

---

* We say, to lessen the dangers of journeys. In fact, although the use of the steam-engine on ships is attended by some danger which has been greatly exaggerated, this is more than compensated by the power of following always an appointed and well-known route, of resisting the force of the winds which would drive the ship towards the shore, the shoals, or the rocks.

† We use here the expression motive power to express the useful effect that a motor is capable of producing. This effect can always be likened to the elevation of a weight to a certain height. It has, as we know, as a measure, the product of the weight multiplied by the height to which it is raised.

example, would not present in this respect great advantages. We propose now to submit these questions to a deliberate examination.

The phenomenon of the production of motion by heat has not been considered from a sufficiently general point of view. We have considered it only in machines the nature and mode of action of which have not allowed us to take in the whole extent of application of which it is susceptible. In such machines the phenomenon is, in a way, incomplete. It becomes difficult to recognize its principles and study its laws.

In order to consider in the most general way the principle of the production of motion by heat, it must be considered independently of any mechanism or any particular agent. It is necessary to establish principles applicable not only to steam-engines* but to all imaginable heat-engines, whatever the working substance and whatever the method by which it is operated.

Machines which do not receive their motion from heat, those which have for a motor the force of men or of animals, a waterfall, an air current, etc., can be studied even to their smallest details by the mechanical theory. All cases are foreseen, all imaginable movements are referred to these general principles, firmly established, and applicable under all circumstances. This is the character of a complete theory. A similar theory is evidently needed for heat-engines. We shall have it only when the laws of physics shall be extended enough, generalized enough, to make known beforehand all the effects of heat acting in a determined manner on any body.

We will suppose in what follows at least a superficial knowledge of the different parts which compose an ordinary steam-engine; and we consider it unnecessary to explain what are the furnace, boiler, steam-cylinder, piston, condenser, etc.

The production of motion in steam-engines is always accompanied by a circumstance on which we should fix our attention. This circumstance is the re-establishing of equilibrium in the caloric; that is, its passage from a body in which the temperature is more or less elevated, to another in which it is lower. What happens in fact in a steam-engine actually in motion? The caloric developed in the furnace by the effect of the combustion traverses the walls of the boiler, produces steam, and in some way incorporates itself with it.

* We distinguish here the steam-engine from the heat-engine in general. The latter may make use of any agent whatever, of the vapor of water or of any other, to develop the motive power of heat.

The latter carrying it away, takes it first into the cylinder, where it performs some function, and from thence into the condenser, where it is liquefied by contact with the cold water which it encounters there. Then, as a final result, the cold water of the condenser takes possession of the caloric developed by the combustion. It is heated by the intervention of the steam as if it had been placed directly over the furnace. The steam is here only a means of transporting the caloric. It fills the same office as in the heating of baths by steam, except that in this case its motion is rendered useful.

We easily recognize in the operations that we have just described the re-establishment of equilibrium in the caloric, its passage from a more or less heated body to a cooler one. The first of these bodies, in this case, is the heated air of the furnace; the second is the condensing water. The re-establishment of equilibrium of the caloric takes place between them, if not completely, at least partially, for on the one hand the heated air, after having performed its function, having passed round the boiler, goes out through the chimney with a temperature much below that which it had acquired as the effect of combustion; and on the other hand, the water of the condenser, after having liquefied the steam, leaves the machine with a temperature higher than that with which it entered.

The production of motive power is then due in steam-engines not to an actual consumption of caloric, but *to its transportation from a warm body to a cold body*, that is, to its re-establishment of equilibrium—an equilibrium considered as destroyed by any cause whatever, by chemical action such as combustion, or by any other. We shall see shortly that this principle is applicable to any machine set in motion by heat.

According to this principle, the production of heat alone is not sufficient to give birth to the impelling power: it is necessary that there should also be cold; without it, the heat would be useless. And in fact, if we should find about us only bodies as hot as our furnaces, how can we condense steam? What should we do with it if once produced? We should not presume that we might discharge it into the atmosphere, as is done in some engines;* the atmosphere would not receive it. It does receive it under the actual condition of things, only because it fulfils the office of a vast

* Certain engines at high pressure throw the steam out into the atmosphere instead of the condenser. They are used specially in places where it would be difficult to procure a stream of cold water sufficient to produce condensation.

condenser, because it is at a lower temperature; otherwise it would soon become fully charged, or rather would be already saturated.*

Wherever there exists a difference of temperature, wherever it has been possible for the equilibrium of the caloric to be re-established, it is possible to have also the production of impelling power. Steam is a means of realizing this power, but it is not the only one. All substances in nature can be employed for this purpose, all are susceptible of changes of volume, of successive contradictions and dilatations, through the alternation of heat and cold. All are capable of overcoming in their changes of volume certain resistances, and of thus developing the impelling power. A solid body—a metallic bar for example—alternately heated and cooled increases and diminishes in length, and can move bodies fastened to its ends. A liquid alternately heated and cooled increases and diminishes in volume, and can overcome obstacles of greater or less size, opposed to its dilatation. An aeriform fluid is susceptible of considerable change of volume by variations of temperature. If it is enclosed in an expansible space, such as a cylinder provided with a piston, it will produce movements of great extent. Vapors of all substances capable of passing into a gaseous condition, as of alcohol, of mercury, of sulphur, etc., may fulfil the same office as vapor of water. The latter, alternately heated and cooled, would produce motive power in the shape of permanent gases, that is, without ever returning to a liquid state. Most of these substances have been proposed, many even have been tried, although up to this time perhaps without remarkable success.

We have shown that in steam-engines the motive-power is due to a re-establishment of equilibrium in the caloric; this takes place not only for steam-engines, but also for every heat-engine—that is,

* The existence of water in the liquid state here necessarily assumed, since without it the steam-engine could not be fed, supposes the existence of a pressure capable of preventing this water from vaporizing, consequently of a pressure equal or superior to the tension of vapor at that temperature. If such a pressure were not exerted by the atmospheric air, there would be instantly produced a quantity of steam sufficient to give rise to that tension, and it would be necessary always to overcome this pressure in order to throw out the steam from the engines into the new atmosphere. Now this is evidently equivalent to overcoming the tension which the steam retains after its condensation, as effected by ordinary means.

If a very high temperature existed at the surface of our globe, as it seems certain that it exists in its interior, all the waters of the ocean would be in a state of vapor in the atmosphere, and no portion of it would be found in a liquid state.

for every machine of which caloric is the motor. Heat can evidently be a cause of motion only by virtue of the changes of volume or of form which it produces in bodies.

These changes are not caused by uniform temperature, but rather by alternations of heat and cold. Now to heat any substance whatever requires a body warmer than the one to be heated; to cool it requires a cooler body. We supply caloric to the first of these bodies that we may transmit it to the second by means of the intermediary substance. This is to re-establish, or at least to endeavor to re-establish, the equilibrium of the caloric.

It is natural to ask here this curious and important question: Is the motive power of heat invariable in quantity, or does it vary with the agent employed to realize it as the intermediary substance, selected as the subject of action of the heat?

It is clear that this question can be asked only in regard to a given quantity of caloric,* the difference of the temperatures also being given. We take, for example, one body *A* kept at a temperature of 100° and another body *B* kept at a temperature of 0°, and ask what quantity of motive power can be produced by the passage of a given portion of caloric (for example, as much as is necessary to melt a kilogram of ice) from the first of these bodies to the second. We inquire whether this quantity of motive power is necessarily limited, whether it varies with the substance employed to realize it, whether the vapor of water offers in this respect more or less advantage than the vapor of alcohol, of mercury, a permanent gas, or any other substance. We will try to answer these questions, availing ourselves of ideas already established.

We have already remarked upon this self-evident fact, or fact which at least appears evident as soon as we reflect on the changes of volume occasioned by heat: *wherever there exists a difference of temperature, motive power can be produced.* Reciprocally, wherever we can consume this power, it is possible to produce a difference of temperature, it is possible to occasion destruction of equilibrium in the caloric. Are not percussion and the friction of bodies actually means of raising their temperature, of making it reach spontaneously a

* It is considered unnecessary to explain here what is quantity of caloric or quantity of heat (for we employ these two expressions indifferently), or to describe how we measure these quantities by the calorimeter. Nor will we explain what is meant by latent heat, degree of temperature, specific heat, etc. The reader should be familiarized with these terms through the study of the elementary treatises of physics or of chemistry.

higher degree than that of the surrounding bodies, and consequently of producing a destruction of equilibrium in the caloric, where equilibrium previously existed? It is a fact proved by experience, that the temperature of gaseous fluids is raised by compression and lowered by rarefaction. This is a sure method of changing the temperature of bodies, and destroying the equilibrium of the caloric as many times as may be desired with the same substance. The vapor of water employed in an inverse manner to that in which it is used in steam-engines can also be regarded as a means of destroying the equilibrium of the caloric. To be convinced of this we need to observe closely the manner in which motive power is developed by the action of heat on vapor of water. Imagine two bodies *A* and *B*, kept each at a constant temperature, that of *A* being higher than that of *B*. These two bodies, to which we can give or from which we can remove the heat without causing their temperatures to vary, exercise the functions of two unlimited reservoirs of caloric. We will call the first the furnace and the second the refrigerator.

If we wish to produce motive power by carrying a certain quantity of heat from the body *A* to the body *B* we shall proceed as follows:*

(1) To borrow caloric from the body *A* to make steam with it—that is, to make this body fulfil the function of a furnace, or rather of the metal composing the boiler in ordinary engines—we here assume that the steam is produced at the same temperature as the body *A*.

(2) The steam having been received in a space capable of expansion, such as a cylinder furnished with a piston, to increase the volume of this space, and consequently also that of the steam. Thus rarefied, the temperature will fall spontaneously, as occurs with all elastic fluids; admit that the rarefaction may be continued to the point where the temperature becomes precisely that of the body *B*.

(3) To condense the steam by putting it in contact with the body *B*, and at the same time exerting on it a constant pressure until it is entirely liquefied. The body *B* fills here the place of the injection-water in ordinary engines, with this difference, that it condenses the vapor without mingling with it, and without changing its own temperature.†

* [This is only a sketch and Carnot accidentally leaves the cycle incomplete. E. M.]

† We may perhaps wonder here that the body *B* being at the same temperature as the steam is able to condense it. Doubtless this is not strictly possible, but the slightest difference of temperature will determine the condensation, which suffices

The operations which we have just described might have been performed in an inverse direction and order. There is nothing to prevent forming vapor with the caloric of the body *B*, and at the temperature of that body, compressing it in such a way as to make it acquire the temperature of the body *A*, finally condensing it by contact with this latter body, and continuing the compression to complete liquefaction.

By our first operations there would have been at the same time production of motive power and transfer of caloric from the body *A* to the body *B*. By the inverse operations there is at the same time expenditure of motive power and return of caloric from the body *B* to the body *A*. But if we have acted in each case on the same quantity of vapor, if there is produced no loss either of motive power or caloric, the quantity of motive power produced in the first place will be equal to that which would have been expended in the second, and the quantity of caloric passed in the first case from the body *A* to the body *B* would be equal to the quantity which passes back again in the second from the body *B* to the body *A*; so that an indefinite number of alternative operations of this sort could be carried on without in the end having either produced motive power or transferred caloric from one body to the other.

Now if there existed any means of using heat preferable to those which we have employed, that is, if it were possible by any method whatever to make the caloric produce a quantity of motive power greater than we have made it produce by our first series of operations, it would suffice to divert a portion of this power in order by the method just indicated to make the caloric of the body *B* return

---

to establish the justice of our reasoning. It is thus that, in the differential calculus, it is sufficient that we can conceive the neglected quantities indefinitely reducible in proportion to the quantities retained in the equations, to make certain of the exact result.

The body *B* condenses the steam without changing its own temperature—this results from our supposition. We have admitted that this body may be maintained at a constant temperature. We take away the caloric as the steam furnishes it. This is the condition in which the metal of the condenser is found when the liquefaction of the steam is accomplished by applying cold water externally, as was formerly done in several engines. Similarly, the water of a reservoir can be maintained at a constant level if the liquid flows out at one side as it flows in at the other.

One could even conceive the bodies *A* and *B* maintaining the same temperature, although they might lose or gain certain quantities of heat. If, for example, the body *A* were a mass of steam ready to become liquid, and the body *B* a mass of ice ready to melt, these bodies might, as we know, furnish or receive caloric without thermometric change.

to the body *A* from the refrigerator to the furnace, to restore the initial conditions, and thus to be ready to commence again an operation precisely similar to the former, and so on: this would be not only perpetual motion, but an unlimited creation of motive power without consumption either of caloric or of any other agent whatever. Such a creation is entirely contrary to ideas now accepted, to the laws of mechanics and of sound physics. It is inadmissible.* We should then conclude that *the maximum of motive power resulting from the employment of steam is also the maximum of motive power realizable by any means whatever.* We will soon give a second more vigorous demonstration of this theory. This should be considered only as an approximation. (See page 15.)

We have a right to ask, in regard to the proposition just enunciated, the following questions: What is the sense of the word *maximum* here? By what sign can it be known that this maximum is attained? By what sign can it be known whether the steam is employed to greatest possible advantage in the production of motive power?

Since every re-establishment of equilibrium in the caloric may be the cause of the production of motive power, every re-establish-

---

* The objection may perhaps be raised here, that perpetual motion, demonstrated to be impossible by mechanical action alone, may possibly not be so if the power either of heat or electricity be exerted; but is it possible to conceive the phenomena of heat and electricity as due to anything else than some kind of motion of the body, and as such should they not be subjected to the general laws of mechanics? Do we not know besides, *à posteriori*, that all the attempts made to produce perpetual motion by any means whatever have been fruitless?—that we have never succeeded in producing a motion veritably perpetual, that is, a motion which will continue forever without alteration in the bodies set to work to accomplish it? The electro-motor apparatus (the pile of Volta) has sometimes been regarded as capable of producing perpetual motion; attempts have been made to realize this idea by constructing dry piles said to be unchangeable; but however it has been done, the apparatus has always exhibited sensible deteriorations when its action has been sustained for a time with any energy.

The general and philosophic acceptation of the words *perpetual motion* should include not only a motion susceptible of indefinitely continuing itself after a first impulse received, but the action of an apparatus, of any construction whatever, capable of creating motive power in unlimited quantity, capable of starting from rest all the bodies of nature if they should be found in that condition, of overcoming their inertia; capable, finally, of finding in itself the forces necessary to move the whole universe, to prolong, to accelerate incessantly, its motion. Such would be a veritable creation of motive power. If this were a possibility, it would be useless to seek in currents of air and water or in combustibles this motive power. We should have at our disposal an inexhaustible source upon which we could draw ta will.

ment of equilibrium which shall be accomplished without production of this power should be considered as an actual loss. Now, very little reflection would show that all change of temperature which is not due to a change of volume of the bodies can be only a useless re-establishment of equilibrium in the caloric.* The necessary condition of the maximum is, then, *that in the bodies employed to realize the motive power of heat there should not occur any change of temperature which may not be due to a change of volume.* Reciprocally, every time that this condition is fulfilled the maximum will be attained. This principle should never be lost sight of in the construction of heat-engines; it is its fundamental basis. If it cannot be strictly observed, it should at least be departed from as little as possible.

Every change of temperature which is not due to a change of volume or to chemical action (an action that we provisionally suppose not to occur here) is necessarily due to the direct passage of the caloric from a more or less heated body to a colder body. This passage occurs mainly by the contact of bodies of different temperatures; hence such contact should be avoided as much as possible. It cannot probably be avoided entirely, but it should at least be so managed that the bodies brought in contact with each other differ as little as possible in temperature. When we just now supposed, in our demonstration, the caloric of the body *A* employed to form steam, this steam was considered as generated at the temperature of the body *A*; thus the contact took place only between bodies of equal temperatures; the change of temperature occurring afterwards in the steam was due to dilatation, consequently to a change of volume. Finally, condensation took place also without contact of bodies of different temperatures. It occurred while exerting a constant pressure on the steam brought in contact with the body *B* of the same temperature as itself. The conditions for a maximum are thus found to be fulfilled. In reality the operation cannot proceed exactly as we have assumed. To determine the passage of caloric from one body to another, it is necessary that there should be an excess of temperature in the first, but this excess may be supposed as slight as we please. We can regard it as insensible in theory, without thereby destroying the exactness of the arguments.

---

* We assume here no chemical action between the bodies employed to realize the motive power of heat. The chemical action which takes place in the furnace is, in some sort, a preliminary action—an operation destined not to produce immediately motive power, but to destroy the equilibrium of the caloric, to produce a difference of temperature which may finally give rise to motion.

A more substantial objection may be made to our demonstration, thus: When we borrow caloric from the body *A* to produce steam, and when this steam is afterwards condensed by its contact with the body *B*, the water used to form it, and which we considered at first as being of the temperature of the body *A*, is found at the close of the operation at the temperature of the body *B*. It has become cool. If we wish to begin again an operation similar to the first, if we wish to develop a new quantity of motive power with the same instrument, with the same steam, it is necessary first to re-establish the original condition—to restore the water to the original temperature. This can undoubtedly be done by at once putting it again in contact with the body *A*; but there is then contact between bodies of different temperatures, and loss of motive power.* It would be impossible to execute the inverse operation, that is, to return to the body *A* the caloric employed to raise the temperature of the liquid.

This difficulty may be removed by supposing the difference of temperature between the body *A* and the body *B* indefinitely small. The quantity of heat necessary to raise the liquid to its former temperature will be also indefinitely small and unimportant relatively to that which is necessary to produce steam—a quantity always limited.

The proposition found elsewhere demonstrated for the case in which the difference between the temperatures of the two bodies is indefinitely small, may be easily extended to the general case. In fact, if it operated to produce motive power by the passage of caloric from the body *A* to the body *Z*, the temperature of this latter body being very different from that of the former, we should imagine a series of bodies *B*, *C*, *D* . . . of temperatures intermediate between those of the bodies *A*, *Z*, and selected so that the differences from *A* to *B*, from *B* to *C*, etc., may all be indefinitely small. The caloric coming from *A* would not arrive at *Z* till after it had passed through

* This kind of loss is found in all steam-engines. In fact, the water destined to feed the boiler is always cooler than the water which it already contains. There occurs between them a useless re-establishment of equilibrium of caloric. We are easily convinced, *à posteriori*, that this re-establishment of equilibrium causes a loss of motive power if we reflect that it would have been possible to previously heat the feed-water by using it as condensing-water in a small accessory engine, when the steam drawn from the large boiler might have been used, and where the condensation might be produced at a temperature intermediate between that of the boiler and that of the principal condenser. The power produced by the small engine would have cost no loss of heat, since all that which had been used would have returned into the boiler with the water of condensation.

the bodies B, C, D, etc., and after having developed in each of these stages maximum motive power. The inverse operations would here be entirely possible, and the reasoning of page 11 would be strictly applicable.

According to established principles at the present time, we can compare with sufficient accuracy the motive power of heat to that of a waterfall. Each has a maximum that we cannot exceed, whatever may be, on the one hand, the machine which is acted upon by the water, and whatever, on the other hand, the substance acted upon by the heat. The motive power of a waterfall depends on its height and on the quantity of the liquid; the motive power of heat depends also on the quantity of caloric used, and on what may be termed, on what in fact we will call, the *height of its fall*,* that is to say, the difference of temperature of the bodies between which the exchange of caloric is made. In the waterfall the motive power is exactly proportional to the difference of level between the higher and lower reservoirs. In the fall of caloric the motive power undoubtedly increases with the difference of temperature between the warm and the cold bodies; but we do not know whether it is proportional to this difference. We do not know, for example, whether the fall of caloric from 100 to 50 degrees furnishes more or less motive power than the fall of this same caloric from 50 to zero. It is a question which we propose to examine hereafter.

We shall give here a second demonstration of the fundamental proposition enunciated on page 12, and present this proposition under a more general form than the one already given.

When a gaseous fluid is rapidly compressed its temperature rises. It falls, on the contrary, when it is rapidly dilated. This is one of the facts best demonstrated by experiment. We will take it for the basis of our demonstration.†

---

* The matter here dealt with being entirely new, we are obliged to employ expressions not in use as yet, and which perhaps are less clear than is desirable.

† The experimental facts which best prove the change of temperature of gases by compression or dilatation are the following:

(1) The fall of the thermometer placed under the receiver of a pneumatic machine in which a vacuum has been produced. This fall is very sensible on the Bréguet thermometer: it may exceed 40° or 50°. The mist which forms in this case seems to be due to the condensation of the watery vapor caused by the cooling of the air.

(2) The igniting of German tinder in the so-called pneumatic tinderboxes; which are, as we know, little pump-chambers in which the air is rapidly compressed.

If, when the temperature of a gas has been raised by compression, we wish to reduce it to its former temperature without subjecting its volume to new changes, some of its caloric must be removed. This caloric might have been removed in proportion as pressure was applied, so that the temperature of the gas would remain constant. Similarly, if the gas is rarefied we can avoid lowering the temperature by supplying it with a certain quantity of caloric. Let us call the caloric employed at such times, when no change of temperature occurs, *caloric due to change of volume*. This denomination does not indicate that the caloric appertains to the volume: it does not appertain to it any more than to pressure, and might as well be called *caloric due to the change of pressure*. We do not know what laws it follows relative to the variations of volume: it is possible that its quantity changes either with the nature of the gas, its density,

---

(3) The fall of a thermometer placed in a space where the air has been first compressed and then allowed to escape by the opening of a cock.

(4) The results of experiments on the velocity of sound. M. de Laplace has shown that, in order to secure results accurately by theory and computation, it is necessary to assume the heating of the air by sudden compression.

The only fact which may be adduced in opposition to the above is an experiment of MM. Gay-Lussac and Welter, described in the *Annales de Chimie et de Physique*. A small opening having been made in a large reservoir of compressed air, and the ball of a thermometer having been introduced into the current of air which passes out through this opening, no sensible fall of the temperature denoted by the thermometer has been observed.

Two explanations of this fact may be given: (1) The striking of the air against the walls of the opening by which it escapes may develop heat in observable quantity. (2) The air which has just touched the ball of the thermometer possibly takes again by its collision with this ball, or rather by the effect of the *détour* which it is forced to make by its rencounter, a density equal to that which it had in the receiver—much as the water of a current rises against a fixed obstacle, above its level.

The change of temperature occasioned in the gas by the change of volume may be regarded as one of the most important facts of physics, because of the numerous consequences which it entails, and at the same time as one of the most difficult to illustrate, and to measure by decisive experiments. It seems to present in some respects singular anomalies.

Is it not to the cooling of the air by dilatation that the cold of the higher regions of the atmosphere must be attributed? The reasons given heretofore as an explanation of this cold are entirely insufficient; it has been said that the air of the elevated regions receiving little reflected heat from the earth, and radiating towards celestial space, would lose caloric, and that this is the cause of its cooling; but this explanation is refuted by the fact that, at an equal height, cold reigns with equal and even more intensity on the elevated plains than on the summit of the mountains, or in those portions of the atmosphere distant from the sun.

or its temperature. Experiment has taught us nothing on this subject. It has only shown us that this caloric is developed in greater or less quantity by the compression of the elastic fluids.

This preliminary idea being established, let us imagine an elastic fluid, atmospheric air for example, shut up in a cylindrical vessel, *abcd* (Fig. 1), provided with a movable diaphragm or piston, *cd*. Let there be also two bodies, *A* and *B*, kept each at a constant temperature, that of *A* being higher than that of *B*. Let us picture to ourselves now the series of operations which are to be described:*

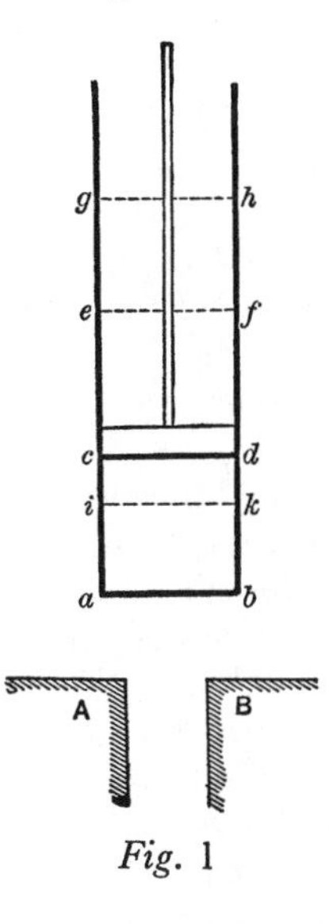

Fig. 1

(1) Contact of the body *A* with the air enclosed in the space *abcd* or with the wall of this space—a wall that we will suppose to transmit the caloric readily. The air becomes by such contact of the same temperature as the body *A*; *cd* is the actual position of the piston.

(2) The piston gradually rises and takes the position *ef*. The body *A* is all the time in contact with the air, which is thus kept at a constant temperature during the rarefaction. The body *A* furnishes the caloric necessary to keep the temperature constant.

(3) The body *A* is removed, and the air is then no longer in contact with any body capable of furnishing it with caloric. The

* ["Caloric" may be taken to mean "entropy." E. M.]

piston meanwhile continues to move, and passes from the position *ef* to the position *gh*. The air is rarefied without receiving caloric, and its temperature falls. Let us imagine that it falls thus till it becomes equal to that of the body *B*; at this instant the piston stops, remaining at the position *gh*.

(4) The air is placed in contact with the body *B*; it is compressed by the return of the piston as it is moved from the position *gh* to the position *cd*. This air remains, however, at a constant temperature because of its contact with the body *B*, to which it yields its caloric.

(5) The body *B* is removed, and the compression of the air is continued, which being then isolated, its temperature rises. The compression is continued till the air acquires the temperature of the body *A*. The piston passes during this time from the position *cd* to the position *ik*.

(6) The air is again placed in contact with the body *A*. The piston returns from the position *ik* to the position *ef*; the temperature remains unchanged.

(7) The step described under number (3) is renewed, then successively the steps (4), (5), (6), (3), (4), (5), (6), (3), (4), (5); and so on.

In these various operations the piston is subject to an effort of greater or less magnitude, exerted by the air enclosed in the cylinder; the elastic force of this air varies as much by reason of the changes in volume as of changes of temperature. But it should be remarked that with equal volumes, that is, for the similar positions of the piston, the temperature is higher during the movements of dilatation than during the movements of compression. During the former the elastic force of the air is found to be greater, and consequently the quantity of motive power produced by the movements of dilatation is more considerable than that consumed to produce the movements of compression. Thus we should obtain an excess of motive power—an excess which we could employ for any purpose whatever. The air, then, has served as a heat-engine; we have, in fact, employed it in the most advantageous manner possible, for no useless re-establishment of equilibrium has been effected in the caloric.

All the above-described operations may be executed in an inverse sense and order. Let us imagine that, after the sixth period, that is to say the piston having arrived at the position *ef*, we cause it to return to the position *ik*, and that at the same time we keep the air in contact with the body *A*. The caloric furnished by this body during the sixth period would return to its source, that is, to the body

A, and the conditions would then become precisely the same as they were at the end of the fifth period. If now we take away the body *A*, and if we cause the piston to move from *ik* to *cd*, the temperature of the air will diminish as many degrees as it increased during the fifth period, and will become that of the body *B*. We may evidently continue a series of operations the inverse of those already described. It is only necessary under the same circumstances to execute for each period a movement of dilatation instead of a movement of compression, and reciprocally.

The result of these first operations has been the production of a certain quantity of motive power and the removal of caloric from the body *A* to the body *B*. The result of the inverse operations is the consumption of the motive power produced and the return of the caloric from the body *B* to the body *A*; so that these two series of operations annul each other, after a fashion, one neutralizing the other.

The impossibility of making the caloric produce a greater quantity of motive power than that which we obtained from it by our first series of operations, is now easily proved. It is demonstrated by reasoning very similar to that employed at page 11; the reasoning will here be even more exact. The air which we have used to develop the motive power is restored at the end of each cycle of operations exactly to the state in which it was at first found, while, as we have already remarked, this would not be precisely the case with the vapor of water.*

We have chosen atmospheric air as the instrument which should develop the motive power of heat, but it is evident that the reasoning would have been the same for all other gaseous substances, and even for all other bodies susceptible of change of temperature through

---

* We tacitly assume in our demonstration, that when a body has experienced any changes, and when after a certain number of transformations it returns to precisely its original state, that is, to that state considered in respect to density, to temperature, to mode of aggregation—let us suppose, I say, that this body is found to contain the same quantitiy of heat that it contained at first, or else that the quantities of heat absorbed or set free in these different transformations are exactly compensated. This fact has never been called in question. It was first admitted without reflection, and verified afterwards in many cases by experiments with the calorimeter. To deny it would be to overthrow the whole theory of heat to which it serves as a basis. For the rest, we may say in passing, the main principles on which the theory of heat rests require the most careful examination. Many experimental facts appear almost inexplicable in the present state of this theory.

successive contractions and dilatations, which comprehends all natural substances, or at least all those which are adapted to realize the motive power of heat. Thus we are led to establish this general proposition:

The motive power of heat is independent of the agents employed to realize it; its quantity is fixed solely by the temperatures of the bodies between which is effected, finally, the transfer of the caloric.

We must understand here that each of the methods of developing motive power attains the perfection of which it is susceptible. This condition is found to be fulfilled if, as we remarked above, there is produced in the body no other change of temperature than that due to change of volume, or, what is the same thing in other words, if there is no contact between bodies of sensibly different temperatures.

Different methods of realizing motive power may be taken, as in the employment of different substances, or in the use of the same substance in two different states—for example, of a gas at two different densities.

This leads us naturally to those interesting researches on the aeriform fluids—researches which lead us also to new results in regard to the motive power of heat, and give us the means of verifying, in some particular cases, the fundamental proposition above stated.*

We readily see that our demonstration would have been simplified by supposing the temperatures of the bodies *A* and *B* to differ very little. Then the movements of the piston being slight during the periods (3) and (5), these periods might have been suppressed without influencing sensibly the production of motive power. A very little change of volume should suffice in fact to produce a very slight change of temperature, and this slight change of volume may be neglected in presence of that of the periods (4) and (6), of which the extent is unlimited.

If we suppress periods (3) and (5), in the series of operations above described, it is reduced to the following:

(1) Contact of the gas confined in *abcd* (Fig. 2) with the body *A*, passage of the piston from *cd* to *ef*.

(2) Removal of the body *A*, contact of the gas confined in *abef* with the body *B*, return of the piston from *ef* to *cd*.

---

* We will suppose, in what follows, the reader to be *au courant* with the later progress of modern physics in regard to gaseous substances and heat.

(3) Removal of the body *B*, contact of the gas with the body *A*, passage of the piston from *cd* to *ef*, that is, repetition of the first period, and so on.

The motive power resulting from the *ensemble* of operations (1) and (2) will evidently be the difference between that which is produced by the expansion of the gas while it is at the temperature of the body *A*, and that which is consumed to compress this gas while it is at the temperature of the body *B*.

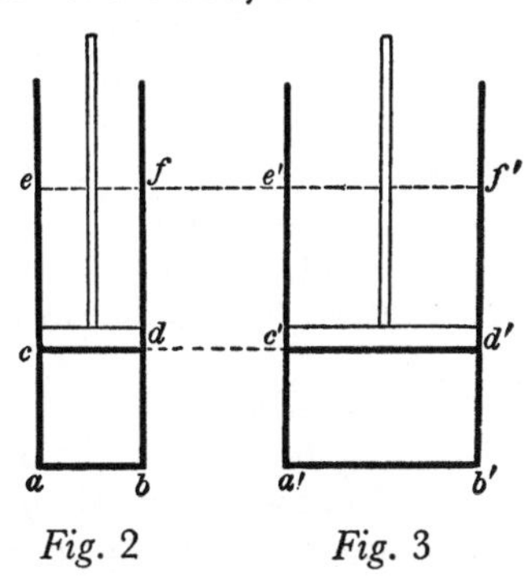

Fig. 2 *Fig.* 3

Let us suppose that operations (1) and (2) be performed on two gases of different chemical natures but under the same pressure—under atmospheric pressure, for example. These two gases will behave exactly alike under the same circumstances, that is, their expansive forces, originally equal, will remain always equal, whatever may be the variations of volume and of temperature, provided these variations are the same in both. This results obviously from the laws of Mariotte and MM. Gay-Lussac and Dalton—laws common to all elastic fluids, and in virtue of which the same relations exist for all these fluids between the volume, the expansive force, and the temperature.

Since two different gases at the same temperature and under the same pressure should behave alike under the same circumstances, if we subjected them both to the operations above described, they should give rise to equal quantities of motive power.

Now this implies, according to the fundamental proposition that we have established, the employment of two equal quantities of caloric; that is, it implies that the quantity of caloric transferred from the body *A* to the body *B* is the same, whichever gas is used.

The quantity of caloric transferred from the body *A* to the body *B* is evidently that which is absorbed by the gas in its expansion of volume, or that which this gas relinquishes during compression. We are led, then, to establish the following proposition:

When a gas passes without change of temperature from one definite volume and pressure to another volume and another pressure equally definite, the quantity of caloric absorbed or relinquished is always the same, whatever may be the nature of the gas chosen as the subject of the experiment.

Take, for example, 1 liter of air at the temperature of 100° and under the pressure of one atmosphere. If we double the volume of this air and wish to maintain it at the temperature of 100°, a certain quantity of heat must be supplied to it. Now this quantity will be precisely the same if, instead of operating on the air, we operate upon carbonic-acid gas, upon nitrogen, upon hydrogen, upon vapor of water or of alcohol, that is, if we double the volume of 1 liter of these gases taken at the temperature of 100° and under atmospheric pressure.

It will be the same thing in the inverse sense if, instead of doubling the volume of gas, we reduce it one half by compression. The quantity of heat that the elastic fluids set free or absorb in their changes of volume has never been measured by any direct experiment, and doubtless such an experiment would be very difficult, but there exists a datum which is very nearly its equivalent. This has been furnished by the theory of sound. It deserves much confidence because of the exactness of the conditions which have led to its establishment. It consists in this:

Atmospheric air should rise one degree Centigrade when by sudden compression it experiences a reduction of volume of $\frac{1}{116}$.*

Experiments on the velocity of sound having been made in air under the pressure of 760 millimeters of mercury and at the temperature of 6°, it is only to these two circumstances that our datum has reference. We will, however, for greater facility, refer it to the temperature 0°, which is nearly the same.

Air compressed $\frac{1}{116}$, and thus heated one degree, differs from air heated directly one degree only in its density. The primitive volume being supposed to be V, the compression of $\frac{1}{116}$ reduces it to $V-\frac{1}{116}V$.

Direct heating under constant pressure should, according to the

---

* M. Poisson, to whom this figure is due, has shown that it accords very well with the result of an experiment of MM. Clément and Désormes on the return of air into a vacuum, or rather, into air slightly rarefied. It also accords very nearly with results found by MM. Gay-Lussac and Welter. (See note, p. 30).

rule of M. Gay-Lussac, increase the volume of air $\frac{1}{267}$ above what it would be at 0°: so the air is, on the one hand, reduced to the volume $V-\frac{1}{116}V$; on the other, it is increased to $V+\frac{1}{267}V$.

The difference between the quantities of heat which the air possesses in both cases is evidently the quantity employed to raise it directly one degree; so then the quantity of heat that the air would absorb in passing from the volume $V-\frac{1}{116}V$ to the volume $V+\frac{1}{267}V$ is equal to that which is required to raise it one degree.

Let us suppose now that, instead of heating one degree the air subjected to a constant pressure and able to dilate freely, we inclose it within an invariable space, and that in this condition we cause it to rise one degree in temperature. The air thus heated one degree will differ from the air compressed $\frac{1}{116}$ only by its $\frac{1}{116}$ greater volume. So then the quantity of heat that the air would set free by a reduction of volume of $\frac{1}{116}$ is equal to that which would be required to raise it one degree Centigrade under constant volume. As the differences between the volumes $V-\frac{1}{116}V$, V, and $V+\frac{1}{267}V$ are small relatively to the volumes themselves, we may regard the quantities of heat absorbed by the air in passing from the first of these volumes to the second, and from the first to the third, as sensibly proportional to the changes of volume. We are then led to the establishment of the following relation:

The quantity of heat necessary to raise one degree air under constant pressure is to the quantity of heat necessary to raise one degree the same air under constant volume, in the ratio of the numbers

$$\frac{1}{116}+\frac{1}{267} \text{ to } \frac{1}{116};$$

or, multiplying both by 116×267, in the ratio of the numbers $267+116$ to 267.

This, then, is the ratio which exists between the capacity of air for heat under constant pressure and its capacity under constant volume. If the first of these two capacities is expressed by unity, the other will be expressed by the number $\frac{267}{267+116}$, or very nearly 0.700; their difference, $1-0.700$ or 0.300, will evidently express the quantity of heat which will produce the increase of volume in the air when it is heated one degree under constant pressure.

According to the law of MM. Gay-Lussac and Dalton, this increase of volume would be the same for all other gases; according to the theorem demonstrated on page 22, the heat absorbed by these equal increases of volume is the same for all the elastic fluids, which leads to the establishment of the following proposition:

The difference between specific heat under constant pressure and specific heat under constant volume is the same for all gases.

It should be remarked here that all the gases are considered as taken under the same pressure, atmospheric pressure for example, and that the specific heats are also measured with reference to the volumes.

It is a very easy matter now for us to prepare a table of the specific heat of gases under constant volume, from the knowledge of their specific heats under constant pressure. Here is the table:

TABLE OF THE SPECIFIC HEAT OF GASES

Names of Gases	Specific Heat under Const. Press.	Specific Heat at Const. Vol.
Atmospheric Air	1.000	0.700
Hydrogen Gas	0.903	0.603
Carbonic Acid	1.258	0.958
Oxygen	0.976	0.676
Nitrogen	1.000	0.700
Protoxide of Nitrogen	1.350	1.050
Olefiant Gas	1.553	1.253
Oxide of Carbon	1.034	0.734

The first column is the result of the direct experiments of MM. Delaroche and Bérard on the specific heat of the gas under atmospheric pressure, and the second column is composed of the numbers of the first diminished by 0.300.

The numbers of the first column and those of the second are here referred to the same unit, to the specific heat of atmospheric air under constant pressure.

The difference between each number of the first column and the corresponding number of the second being constant, the ratio between these numbers should be variable. Thus the ratio between the specific heat of gases under constant pressure and the specific heat at constant volume, varies in different gases.

We have seen that air when it is subjected to a sudden compression of $\frac{1}{116}$ of its volume rises one degree in temperature. The other gases through a similar compression should also rise in temperature. They should rise, but not equally, in inverse ratio with their specific heat at constant volume. In fact, the reduction of volume being by hypothesis always the same, the quantity of heat due to this reduction should likewise be always the same, and consequently should produce an elevation of temperature dependent only on the specific heat acquired by the gas after its compression, and evidently in inverse ratio with this specific heat. Thus we can easily form the table of the elevations of temperature of the different gases for a compression of $\frac{1}{116}$.

TABLE OF THE ELEVATION OF TEMPERATURE

OF

GASES THROUGH THE EFFECT OF COMPRESSION

Names of Gases	Elevation of Temperature for a Reduction of Volume of $\frac{1}{116}$
	°
Atmospheric Air	1.000
Hydrogen Gas	1.160
Carbonic Acid	0.730
Oxygen	1.035
Nitrogen	1.000
Protoxide of Nitrogen	0.667
Olefiant Gas	0.558
Carbonic Oxide	0.955

A second compression of $\frac{1}{116}$ (of the altered volume), as we shall presently see, would also raise the temperature of these gases nearly as much as the first; but it would not be the same with a third, a fourth, a hundredth such compression. The capacity of gases for heat changes with their volume. It is not unlikely that it changes also with the temperature.

We shall now deduce from the general proposition stated on page 20 a second theory, which will serve as a corollary to that just demonstrated.

Let us suppose that the gas enclosed in the cylindrical space *abcd* (Fig. 2) be transported into the space *a'b'c'd'* (Fig. 3) of equal height, but of different base and wider. This gas would increase in volume, would diminish in density and in elastic force, in the inverse ratio of the two volumes *abcd*, *a'b'c'd'*. As to the total pressure exerted in each piston *cd*, *c'd'*, it would be the same from all quarters, for the surface of these pistons is in direct ratio to the volumes.

Let us suppose that we perform on the gas inclosed in *a'b'c'd'* the operations described on page 20, and which were taken as having been performed upon the gas inclosed in *abcd*; that is, let us suppose that we have given to the piston *c'd'* motions equal to those of the piston *cd*, that we have made it occupy successively the positions *c'd'* corresponding to *cd*, and *e'f'* corresponding to *ef*, and that at the same time we have subjected the gas by means of the two bodies *A* and *B* to the same variations of temperature as when it was inclosed in *abcd*. The total effort exercised on the piston would be found to be, in the two cases, always the same at the corresponding instants. This results solely from the law of Mariotte.* In fact, the densities of the two gases maintaining always the same ratio for similar positions of the pistons, and the temperatures being always equal in both, the total pressures exercised on the pistons will always maintain the same ratio to each other. If this ratio is, at any instant whatever, unity, the pressures will always be equal.

As, furthermore, the movements of the two pistons have equal extent, the motive power produced by each will evidently be the same; whence we should conclude, according to the proposition on page 20, that the quantities of heat consumed by each are the same, that is, that there passes from the body *A* to the body *B* the same quantity of heat in both cases.

The heat abstracted from the body *A* and communicated to the body *B*, is simply the heat absorbed during the rarefaction of the gas, and afterwards liberated by its compression. We are therefore led to establish the following theorem:

When an elastic fluid passes without change of temperature from the volume U to the volume V, and when a similar ponderable quantity of the same gas passes at the same temperature from the volume U' to the volume V', if the ratio of U' to V' is found to be the same as the ratio of U to V, the quantities of heat absorbed or disengaged in the two cases will be equal.

---

* The law of Mariotte, which is here made the foundation upon which to establish our demonstration, is one of the best authenticated physical laws. It has served as a basis to many theories verified by experience, and which in turn verify all the laws on which they are founded. We can cite also, as a valuable verification of Mariotte's law and also of that of MM. Gay-Lussac and Dalton, for a great difference of temperature, the experiments of MM. Dulong and Petit. (See *Annales de Chimie et de Physique*, Feb. 1818, vii, p. 122).

The more recent experiments of Davy and Faraday can also be cited.

The theories that we deduce here would not perhaps be exact if applied outside of certain limits either of density or temperature. They should be regarded as true only within the limits in which the laws of Mariotte and of MM. Gay-Lussac and Dalton are themselves proven.

This theorem might also be expressed as follows:

*When a gas varies in volume without change of temperature, the quantities of heat absorbed or liberated by this gas are in arithmetical progression, if the increments or the decrements of volume are found to be in geometrical progression.**

When a liter of air maintained at a temperature of ten degrees is compressed, and when it is reduced to one half a liter, a certain quantity of heat is set free. This quantity will be found always the same if the volume is further reduced from a half liter to a quarter liter, from a quarter liter to an eighth, and so on.

If, instead of compressing the air, we carry it successively to two liters, four liters, eight liters, etc., it will be necessary to supply to it always equal quantities of heat in order to maintain a constant temperature.

This readily accounts for the high temperature attained by air when rapidly compressed. We know that this temperature inflames tinder and even makes air luminous. If, for a moment, we suppose the specific heat of air to be constant, in spite of the changes of volume and temperature, the temperature will increase in arithmetical progression for reduction of volume in geometrical progression.

Starting from this datum, and admitting that one degree of elevation in the temperature corresponds to a compression of $\frac{1}{116}$, we shall readily come to the conclusion that air reduced to $\frac{1}{14}$ of its primitive volume should rise in temperature about 300 degrees, which is sufficient to inflame tinder.†

---

* [In a later footnote Carnot gives the formula for the quantity of heat $e = T(t) \log v$ absorbed in an isothermal expansion from volume 1 to v. $T(t)$ is identical with Clapeyron's $C(t)$ and is in fact proportional to the absolute temperature. E. M.]

† When the volume is reduced $\frac{1}{116}$, that is, when it becomes $\frac{115}{116}$ of what it was at first, the temperature rises one degree. Another reduction of $\frac{1}{116}$ carries the volume to $(\frac{115}{116})^2$, and the temperature should rise another degree. After x similar reductions the volume becomes $(\frac{115}{116})^x$, and the temperature should be raised x degrees. If we suppose $(\frac{115}{116})^x = \frac{1}{14}$, and if we take the logarithms of both, we find

$$x = \text{about } 300^\circ.$$

If we suppose $(\frac{115}{116})^x = \frac{1}{2}$, we find

$$x = 80^\circ;$$

which shows that air compressed one half rises 80°.

All this is subject to the hypothesis that the specific heat of air does not change, although the volume diminishes. But if, for the reasons hereafter given (pp. 29, 31), we regard the specific heat of air compressed one half as reduced in the relation of 700 to 616, the number 80° must be multiplied by $\frac{700}{616}$, which raises it to 90°.

The elevation of temperature ought, evidently, to be still more considerable if the capacity of the air for heat becomes less as its volume diminishes. Now this is probable, and it also seems to follow from the experiments of MM. Delaroche and Bérard on the specific heat of air taken at different densities. (See the memoir in the *Annales de Chimie*, lxxxv, pp. 72, 224.)

The two theorems explained on pp. 22 and 26 suffice for the comparison of the quantities of heat absorbed or set free in the changes of volume of elastic fluids, whatever may be the density and the chemical nature of these fluids, provided always that they be taken and maintained at a certain invariable temperature. But these theories furnish no means of comparing the quantities of heat liberated or absorbed by elastic fluids which change in volume at different temperatures. Thus we are ignorant of what relation exists between the heat relinquished by a liter of air reduced one half, the temperature being kept at zero, and the heat relinquished by the same liter of air reduced one half, the temperature being kept at 100°. The knowledge of this relation is closely connected with that of the specific heat of gases at various temperatures, and to some other data that physics as yet does not supply.

The second of our theorems offers us a means of determining according to what law the specific heat of gases varies with their density.

Let us suppose that the operations described on p. 20, instead of being performed with two bodies, A, B, of temperatures differing indefinitely little, were carried on with two bodies whose temperatures differ by a finite quanity—one degree, for example. In a complete circle of operations the body A furnishes to the elastic fluid a certain quantity of heat, which may be divided into two portions: (1) That which is necessary to maintain the temperature of the fluid constant during dilatation; (2) that which is necessary to restore the temperature of the fluid from that of the body B to that of the body A, when, after having brought back this fluid to its primitive volume, we place it again in contact with the body A. Let us call the first of these quantities a and the second b. The total caloric furnished by the body A will be expressed by $a+b$.

The caloric transmitted by the fluid to the body B may also be divided into two parts: one, b', due to the cooling of the gas by the body B; the other, a', which the gas abandons as a result of its reduction of volume. The sum of these two quantities is $a'+b'$; it should be equal to $a+b$, for, after a complete cycle of operations, the gas is brought back exactly to its primitive state. It has been obliged to give up all the caloric which has first been furnished to it. We have then

$$a+b = a'+b';$$

or rather,

$$a-a' = b'-b.$$

Now, according to the theorem given on page 26, the quantities a and a' are independent of the density of the gas, provided always that the ponderable quantity remains the same and that the variations of volume be proportional to the original volume. The difference $a-a'$ should fulfill the same conditions, and consequently also the difference $b'-b$, which is equal to it. But b' is the caloric necessary to raise the gas enclosed in *abcd* (Fig. 2) one degree; b' is the caloric surrendered by the gas when, enclosed in *abef*, it is cooled one degree. These quantities may serve as a measure for specific heats. We are then led to the establishment of the following proposition:

The change in the specific heat of a gas caused by change of volume depends entirely on the ratio between the original volume and the altered volume. That is, the difference of the specific heats does not depend on the absolute magnitude of the volumes, but only on their ratio.

This proposition might also be differently expressed, thus:

*When a gas increases in volume in geometrical progression, its specific heat increases in arithmetical progression.**

Thus, a being the specific heat of air taken at a given density, and $a+h$ the specific heat for a density one half less, it will be, for a density equal to one quarter, $a+2h$; for a density equal to one eighth, $a+3h$; and so on.

The specific heats are here taken with reference to weight. They are supposed to be taken at an invariable volume, but, as we shall see, they would follow the same law if they were taken under constant pressure.

To what cause is the difference between specific heats at constant volume and at constant pressure really due? To the caloric required to produce in the second case increase of volume. Now,

---

* [The result that the specific heat of a perfect gas varies as $\log v$ was unfortunately thought to be confirmed by an experiment of Delaroche and Bérard cited later. E. M.]

according to the law of Mariotte, increase of volume of a gas should be, for a given change of temperature, a determined fraction of the original volume, a fraction independent of pressure. According to the theorem expressed on page 26, if the ratio between the primitive volume and the altered volume is given, that determines the heat necessary to produce increase of volume. It depends solely on this ratio and on the weight of the gas. We must then conclude that:

The difference between specific heat at constant pressure and specific heat at constant volume is always the same, whatever may be the density of the gas, provided the weight remains the same.

These specific heats both increase accordingly as the density of the gas diminishes, but their difference does not vary.*

Since the difference between the two capacities for heat is constant, if one increases in arithmetical progression the other should follow a similiar progression: thus our law is applicable to specific heats at constant pressure.

We have tacitly assumed the increase of specific heat with that of volume. This increase is indicated by the experiments of MM. Delaroche and Bérard: in fact these physicists have found 0.967 for the specific heat of air under the pressure of 1 meter of mercury (see Memoir already cited), taking for the unit the specific heat of the same weight of air under the pressure of $0^{m}.760$.

According to the law that specific heats follow with relation to pressures, it is only necessary to have observed them in two particular cases to deduce them in all possible cases: it is thus that, making use of the experimental result of MM. Delaroche and Bérard which

---

* MM. Gay-Lussac and Welter have found by direct experiments, cited in the *Mécanique Céleste* and in the *Annales de Chimie et de Physique*, July, 1822, p. 267, that the ratio between the specific heat at constant pressure and the specific heat at constant volume varies very little with the density of the gas. According to what we have just seen, the difference should remain constant, and not the ratio. As, further, the specific heat of gases for a given weight varies very little with the density, it is evident that the ratio itself experiences but slight changes.

The ratio between the specific heat of atmospheric air at constant pressure and at constant volume is, according to MM. Gay-Lussac and Welter, 1.3748, a number almost constant for all pressures, and even for all temperatures. We have come, through other considerations, to the number $\frac{267+116}{267} = 1.44$, which differs from the former $\frac{1}{20}$, and we have used this number to prepare a table of the specific heats of gases at constant volume. So we need not regard this table as very exact, any more than the table given on p. 31. These tables are mainly intended to demonstrate the laws governing specific heats of aeriform fluids.

has just been given, we have prepared the following table of the specific heat of air under different pressures:

SPECIFIC HEAT OF AIR

Pressure in Atmospheres	Specific Heat, that of Air under Atmospheric Pressure being 1	Pressure in Atmospheres	Specific Heat, that of Air under Atmospheric Pressure being 1
$\frac{1}{1024}$	1.840	1	1.000
$\frac{1}{512}$	1.756	2	0.916
$\frac{1}{256}$	1.672	4	0.832
$\frac{1}{128}$	1.588	8	0.748
$\frac{1}{64}$	1.504	16	0.664
$\frac{1}{32}$	1.420	32	0.580
$\frac{1}{16}$	1.336	64	0.496
$\frac{1}{8}$	1.252	128	0.412
$\frac{1}{4}$	1.165	256	0.328
$\frac{1}{2}$	1.084	512	0.244
1	1.000	1024	0.160

The first column is, as we see, a geometrical progression, and the second an arithmetical progression.

We have carried out the table to the extremes of compression and rarefaction. It may be believed that air would be liquefied before acquiring a density 1024 times its normal density, that is, before becoming more dense than water. The specific heat would become zero and even negative on extending the table beyond the last term. We think, furthermore, that the figures of the second column here decrease too rapidly. The experiments which serve as a basis for our calculation have been made within too contracted limits for us to expect great exactness in the figures which we have obtained, especially in the outside numbers.

Since we know, on the one hand, the law according to which heat is disengaged in the compression of gases, and on the other, the law according to which specific heat varies with volume, it will be easy for us to calculate the increase of temperature of a gas that has been compressed without being allowed to lose heat. In fact, the compression may be considered as composed of two successive operations: (1) compression at a constant temperature; (2) restoration of the caloric emitted. The temperature will rise through the second operation in inverse ratio with the specific heat acquired by the gas after the reduction of volume—specific heat that we are able to calculate by means of the law demonstrated above. The heat set free by compression, according to the theorem of page 27, ought to be represented by an expression of the form

$$s = A + B \log v,$$

s being this heat, v the volume of the gas after compression, A and B arbitrary constants dependent on the primitive volume of the gas, on its pressure, and on the units chosen.

The specific heat varying with the volume according to the law just demonstrated, should be represented by an expression of the form

$$z = A' + B' \log v,$$

A' and B' being different arbitrary constants from A and B.

The increase of temperature acquired by the gas, as the effect of compression, is proportional to the ratio $\frac{s}{z}$ or to the relation $\frac{A + B \log v}{A' + B' \log v}$. It can be represented by this ratio itself; thus, calling it t, we shall have

$$t = \frac{A + B \log v}{A' + B' \log v}.$$

If the original volume of the gas is 1, and the original temperature zero, we shall have at the same time $t = 0$, $\log v = 0$, whence $A = 0$; t will then express not only the increase of temperature, but the temperature itself above the thermometric zero.

We need not consider the formula that we have just given as applicable to very great changes in the volume of gases. We have regarded the elevation of temperature as being in inverse ratio to the specific heat; which tacitly supposes the specific heat to be constant at all temperatures. Great changes of volume lead to great changes of temperature in the gas, and nothing proves the constancy of specific heat at different temperatures, especially at temperatures widely separated. This constancy is only an hypothesis admitted for gases by analogy, to a certain extent verified for solid bodies and liquids throughout a part of the thermometric scale, but of which the experiments of MM. Dulong and Petit have shown the inaccuracy when it is desirable to extend it to temperatures far above 100°.*

* We see no reason for admitting, *à priori*, the constancy of the specific heat of bodies at different temperatures, that is, to admit that equal quantities of heat will produce equal increments of temperature, when this body changes neither its

According to a law of MM. Clément and Désormes,* a law established by direct experiment, the vapor of water, under whatever pressure it may be formed, contains always, at equal weights, the same quantity of heat; which leads to the assertion that steam, compressed or expanded mechanically without loss of heat, will always be found in a saturated state if it was so produced in the first place. The vapor of water so made may then be regarded as a permanent gas, and should observe all the laws of one. Consequently the formula

$$t = \frac{A + B \log v}{A' + B' \log v}$$

should be applicable to it, and be found to accord with the table of tensions derived from the direct experiments of M. Dalton.

We may be assured, in fact, that our formula, with a convenient determination of arbitrary constants, represents very closely the results of experiment. The slight irregularities which we find therein do not exceed what we might reasonably attribute to errors of observation.†

---

state nor its density; when, for example, it is an elastic fluid enclosed in a fixed space. Direct experiments on solid and liquid bodies have proved that between zero and 100°, equal increments in the quantities of heat would produce nearly equal increments of temperature. But the more recent experiments of MM. Dulong and Petit (see *Annales de Chimie et de Physique*, February, March, and April, 1818) have shown that this correspondence no longer continues at temperatures much above 100°, whether these temperatures be measured on the mercury thermometer or on the air thermometer.

Not only do the specific heats not remain the same at different temperatures, but, also, they do not preserve the same ratios among themselves, so that no thermometric scale could establish the constancy of all the specific heats. It would have been interesting to prove whether the same irregularities exist for gaseous substances, but such experiments presented almost insurmountable difficulties.

The irregularities of specific heats of solid bodies might have been attributed, it would seem, to the latent heat employed to produce a beginning of fusion—a softening which occurs in most bodies long before complete fusion. We might support this opinion by the following statement: According to the experiments of MM. Dulong and Petit, the increase of specific heat with the temperature is more rapid in solids than in liquids, although the latter possess considerably more dilatability. The cause of irregularity just referred to, if it is real, would disappear entirely in gases.

* [In English this is called Watt's law, that the enthalpy of saturated steam is constant at all temperatures. E. M.]

† In order to determine the arbitrary constants A, B, A', B', in accordance with the results in M. Dalton's table, we must begin by computing the volume of the vapor as determined by its pressure and temperature—a result which is easily

accomplished by reference to the laws of Mariotte and Gay-Lussac, the weight of the vapor being fixed.

The volume will be given by the equation

$$v = c\,\frac{267+t}{p},$$

in which v is this volume, t the temperature, p the pressure, and c a constant quantity depending on the weight of the vapor and on the units chosen. We give here the table of the volumes occupied by a gram of vapor formed at different temperatures, and consequently under different pressures.

t or degrees Centigrade	p or tension of the vapor expressed in millimeters of mercury	v or volume of a gram of vapor expressed in liters
°		
0	5.060	185.0
20	17.32	58.2
40	53.00	20.4
60	144.6	7.96
80	352.1	3.47
100	760.0	1.70

The first two columns of this table are taken from the *Traité de Physique* of M. Biot (vol. i., p. 272 and 531). The third is calculated by means of the above formula, and in accordance with the result of experiment, indicating that water vaporized under atmospheric pressure occupies a space 1700 times as great as in the liquid state.

By using three numbers of the first column and three corresponding numbers of the third column, we can easily determine the constants of our equation

$$t = \frac{A+B\log v}{A'+B'\log v}.$$

We will not enter into the details of the calculation necessary to determine these quantities. It is sufficient to say that the following values,

$$A = 2268, \qquad A' = 19.64,$$
$$B = -1000, \qquad B' = 3.30,$$

satisfy fairly well the prescribed conditions, so that the equation

$$t = \frac{2268-1000\log v}{19.64+3.30\log v}$$

expresses very nearly the relation which exists between the volume of the vapor and its temperature. We may remark here that the quantity B' is positive and very small, which tends to confirm this proposition—that the specific heat of an elastic fluid increases with the volume, but follows a slow progression.

We will return, however, to our principal subject, from which we have wandered too far—the motive power of heat.

We have shown that the quantity of motive power developed by the transfer of caloric from one body to another depends essentially upon the temperature of the two bodies, but we have not shown the relation between these temperatures and the quantities of motive power produced. It would at first seem natural enough to suppose that for equal differences of temperature the quantities of motive power produced are equal; that is, for example, the passage of a given quantity of caloric from a body, *A*, maintained at 100°, to a body, *B*, maintained at 50°, should give rise to a quantity of motive power equal to that which would be developed by the transfer of the same caloric from a body, *B*, at 50°, to a body, *C*, at zero. Such a law would doubtless be very remarkable, but we do not see sufficient reason for admitting it *à priori*. We will investigate its reality by exact reasoning.

Let us imagine that the operations described on p. 20 be conducted successively on two quantities of atmospheric air equal in weight and volume, but taken at different temperatures. Let us suppose, further, the differences of temperature between the bodies *A* and *B* equal, so these bodies would have for example, in one of these cases, the temperatures 100° and $100° - h$ (h being indefinitely small), and in the other 1° and $1° - h$. The quantity of motive power produced is, in each case, the difference between that which the gas supplies by its dilatation and that which must be expended to restore its primitive volume. Now this difference is the same in both cases, as any one can prove by simple reasoning, which it seems unnecessary to give here in detail; hence the motive power produced is the same.

Let us now compare the quantities of heat employed in the two cases. In the first, the quantity of heat employed is that which the body *A* furnishes to the air to maintain it at the temperature of 100° during its expansion. In the second, it is the quantity of heat which this same body should furnish to it, to keep its temperature at one degree during an exactly similar change of volume. If these two quantities of heat were equal, there would evidently result the law that we have already assumed. But nothing proves that it is so, and we shall find that these quantities are not equal.

The air that we shall first consider as occupying the space *abcd* (Fig. 2), and having 1 degree of temperature, can be made to occupy the space *abef*, and to acquire the temperature of 100 degrees by two different means:

(1) We may heat it without changing its volume, then expand it, keeping its temperature constant.

(2) We may begin by expanding it, maintaining the temperature constant, then heat it, when it has acquired its greater volume.

Let a and b be the quantities of heat employed successively in the first of the two operations, and let b' and a' be the quantities of heat employed successively in the second. As the final result of these two operations is the same, the quantities of heat employed in both should be equal. We have then

$$a+b = a'+b',$$

whence

$$a'-a = b-b'.$$

a' is the quantity of heat required to cause the gas to rise from 1° to 100° when it occupies the space *abef*.

a is the quantity of heat required to cause the gas to rise from 1° to 100° when it occupies the space *abcd*.

The density of the air is less in the first than in the second case, and according to the experiments of MM. Delaroche and Bérard, already cited on page 30, its capacity for heat should be a little greater.

The quantity a' being found to be greater than the quantity a, b should be greater than b'. Consequently, generalizing the proposition, we should say:

The quantity of heat due to the change of volume of a gas is greater as the temperature is higher.

Thus, for example, more caloric is necessary to maintain at 100° the temperature of a certain quantity of air the volume of which is doubled, than to maintain at 1° the temperature of this same air during a dilatation exactly equal.

These unequal quantities of heat would produce, however, as we have seen, equal quantities of motive power for equal fall of caloric taken at different heights on the thermometric scale; whence we draw the following conclusion:

*The fall of caloric produces more motive power at inferior than at superior temperatures.**

---

* [Carnot later deduces *either* that the motive power for a given temperature interval varies quite strongly with temperature *or* that it is constant—an incorrect result based ultimately on the wrong measurements of specific heats under pressure. E. M.]

Thus a given quantity of heat will develop more motive power in passing from a body kept at 1 degree to another maintained at zero, than if these two bodies were at the temperature of 101° and 100°.

The difference, however, should be very slight. It would be nothing if the capacity of the air for heat remained constant, in spite of changes of density. According to the experiments of MM. Delaroche and Bérard, this capacity varies little—so little even, that the differences noticed might strictly have been attributed to errors of observation or to some circumstances of which we have failed to take account.

We are not prepared to determine precisely, with no more experimental data than we now possess, the law according to which the motive power of heat varies at different points on the thermometric scale. This law is intimately connected with that of the variations of the specific heat of gases at different temperatures—a law which experiment has not yet made known to us with sufficient exactness.*

---

* Were we to admit the constancy of the specific heat of a gas when its volume does not change, but when its temperature varies, analysis would show a relation between the motive power and the thermometric degree. We will show how this is, and this will also give us occasion to show how some of the propositions established above should be expressed in algebraic language.

Let r be the quantity of motive power produced by the expansion of a given quantity of air passing from the volume of one liter to the volume of v liters under constant temperature. If v increases by the infinitely small quantity dv, r will increase by the quantity dr, which, according to the nature of motive power, will be equal to the increase dv of volume multiplied by the expansive force which the elastic fluid then possesses; let p be this expansive force. We should have the equation

$$dr = pdv. \qquad \ldots\ldots\ldots \qquad (1)$$

Let us suppose the constant temperature under which the dilatation takes place equal to t degrees Centigrade. If we call q the elastic force of the air occupying the volume 1 liter at the same temperature t, we shall have, according to the law of Mariotte,

$$\frac{v}{1} = \frac{q}{p}, \text{ whence } p = \frac{q}{v}.$$

If now P is the elastic force of this same air at the constant volume 1, but at the temperature zero, we shall have, according to the rule of M. Gay-Lussac,

$$q = P + P\frac{t}{267} = \frac{P}{267}(267+t);$$

whence

$$\frac{q}{v} = p = \frac{P}{267}\frac{267+t}{v}.$$

If, to abridge, we call N the quantity $\frac{P}{267}$, the equation would become

$$p = N\frac{t+267}{v};$$

whence we deduce, according to equation (1),

$$dr = N\frac{t+267}{v}\,dv.$$

Regarding t as constant, and taking the integral of the two members, we shall have

$$r = N(t+267)\log v+C.$$

If, we suppose $r=0$ when $v=1$, we shall have $C=0$; whence

$$r = N(t+267)\log v. \qquad (2)$$

This is the motive power produced by the expansion of the air which, under the temperature t, has passed from the volume 1 to the volume v. If instead of working at the temperature t we work in precisely the same manner at the temperature $t+dt$, the power developed will be

$$r+\delta r = N(t+dt+267)\log v.$$

Subtracting equation (2), we have

$$\delta r = N\log v dt. \qquad (3)$$

Let e be the quantity of heat employed to maintain the temperature of the gas constant during its dilatation. According to the reasoning of page 21, δr will be the power developed by the fall of the quantity e of heat from the degree $t+td$ to the degree t. If we call u the motive power developed by the fall of unity of heat from the degree t to the degree zero, as, according to the general principle established, page 20, this quantity u ought to depend solely on t, it could be represented by the function Ft, whence $u=Ft$.

When t is increased it becomes $t+td$, u becomes $u+du$; whence

$$u+du = F(t+dt).$$

Subtracting the preceding equation, we have

$$du = F(t+dt)-Ft = F'tdt.$$

This is evidently the quantity of motive power produced by the fall of unity of heat from the temperature $t+dt$ to the temperature t.

If the quantity of heat instead of being a unit had been e, its motive power produced would have had for its value

$$edu = eF'tdt. \qquad (4)$$

But edu is the same thing as δr; both are the power developed by the fall of the quantity e of heat from the temperature $t+dt$ to the temperature t; consequently,

$$edu = \delta r,$$

and from equations (3), (4),

$$eF'tdt = N\log v dt;$$

or, dividing by $F'tdt$,

$$e = \frac{N}{F't} \log v = T \log v,$$

calling T the fraction $\frac{N}{F't}$ which is a function of t only. The equation

$$e = T \log v$$

is the analytical expression of the law stated on pp. 26, 27. It is common to all gases, since the laws of which we have made use are common to all.

If we call s the quantity of heat necessary to change the air that we have employed from the volume 1 and from the temperature zero to the volume v and to the temperature t, the difference between s and e will be the quantity of heat required to bring the air at the volume 1 from zero to t. This quantity depends on t alone; we will call it U. It will be some function of t. We shall have

$$s = e + U = T \log v + U.$$

If we differentiate this equation with relation to t alone, and if we represent by T' and U', the differential coefficients of T and U, we shall get

$$\frac{ds}{dt} = T' \log v + U'; \qquad \qquad (5)$$

$\frac{ds}{dt}$ is simply the specific heat of the gas under constant volume, and our equation (5) is the analytical expression of the law stated on page 29.

If we suppose the specific heat constant at all temperatures (hypothesis discussed above, page 32), the quantity $\frac{ds}{dt}$ will be independent of t; and in order to satisfy equation (5) for two particular values of v, it will be necessary that T' and U' be independent of t; we shall then have $T' = C$, a constant quantity. Multiplying T' and C by dt, and taking the integral of both, we find

$$T = Ct + C_1;$$

but as $T = \frac{N}{F't}$, we have

$$F't = \frac{N}{T} = \frac{N}{Ct + C_1}.$$

Multiplying both by dt and integrating, we have

$$F = \frac{N}{C} \log (Ct + C_1) + C_2;$$

or changing arbitrary constants, and remarking further that Ft is 0 when $t = 0°$,

$$Ft = A \log \left(1 + \frac{t}{B}\right). \qquad \qquad (6)$$

The nature of the function Ft would be thus determined, and we would thus be able to estimate the motive power developed by any fall of heat. But this latter conclusion is founded on the hypothesis of the constancy of the specific heat of a gas which does not change in volume—an hypothesis which has not yet been sufficiently verified by experiment. Until there is fresh proof, our equation (6) can be admitted only throughout a limited portion of the thermometric scale.

We will endeavor now to estimate exactly the motive power of heat, and in order to verify our fundamental proposition, in order to determine whether the agent used to realize the motive power is really unimportant relatively to the quantity of this power, we will select several of them successively: atmospheric air, vapor of water, vapor of alcohol.

Let us suppose that we take first atmospheric air. The operation will proceed according to the method indicated on page 20. We will make the following hypotheses: The air is taken under atmospheric pressure. The temperature of the body *A* is $\frac{1}{1000}$ of a degree above zero, that of the body *B* is zero. The difference is, as we see, very slight—a necessary condition here.

The increase of volume given to the air in our operation will be $\frac{1}{116}+\frac{1}{267}$ of the primitive volume; this is a very slight increase, absolutely speaking, but great relatively to the difference of temperature between the bodies *A* and *B*.

The motive power developed by the whole of the two operations described (page 20) will be very nearly proportional to the increase of volume and to the difference between the two pressures exerted by the air, when it is found at the temperature $0^\circ.001$ and zero.

In equation (5), the 1st term represents, as we have remarked, the specific heat of the air occupying the volume v. Experiments having shown that this heat varies little in spite of the quite considerable changes of volume, it is necessary that the coefficient T' of log v should be a very small quantity. If we consider it nothing, and, after having multiplied by dt the equation

$$T' = 0,$$

we take the integral of it, we find

$$T = C, \text{ constant quantity;}$$

but

$$T = \frac{N}{F't},$$

whence

$$F't = \frac{N}{T} = \frac{N}{C} = A;$$

whence we deduce finally, by a second integration,

$$Ft = At + B.$$

As $Ft=0$ when $t=0$, B is 0; thus

$$Ft = At;$$

that is, the motive power produced would be found to be exactly proportional to the fall of the caloric. This is the analytical translation of what was stated on page 35.

This difference is, according to the law of M. Gay-Lussac, $\frac{1}{267000}$ of the elastic force of the gas, or very nearly $\frac{1}{267000}$ of the atmospheric pressure.

The atmospheric pressure balances at 10.40 meters head of water; $\frac{1}{267000}$ of this pressure equals $\frac{1}{267000} \times 10^{m}.40$ of head of water.

As to the increase of volume, it is, by supposition, $\frac{1}{116}+\frac{1}{267}$ of the original volume, that is, of the volume occupied by one kilogram of air at zero, a volume equal to $0^{mc}.77$, allowing for the specific weight of the air. So then the product,

$$\left(\frac{1}{116}+\frac{1}{267}\right) \times 0.77 \times \frac{1}{267000} \times 10.40,$$

will express the motive power developed. This power is estimated here in cubic meters of water raised one meter.

If we carry out the indicated multiplications, we find the value of the product to be 0.000000372.

Let us endeavor now to estimate the quantity of heat employed to give this result; that is, the quantity of heat passed from the body *A* to the body *B*.

The body *A* furnishes:

(1) The heat required to carry the temperature of one kilogram of air from zero to $0^{\circ}{\cdot}001$;

(2) The quantity necessary to maintain at this temperature the temperature of the air when it experiences a dilatation of

$$\frac{1}{116}+\frac{1}{267}.$$

The first of these quantities of heat being very small in comparison with the second, we may disregard it. The second is, according to the reasoning on page 23, equal to that which would be necessary to increase one degree the temperature of one kilogram of air subjected to atmospheric pressure.

According to the experiments of MM. Delaroche and Bérard on the specific heat of gases, that of air is, for equal weights, 0.267 that of water. If, then, we take for the unit of heat the quantity necessary to raise 1 kilogram of water 1 degree, that which will be required to raise 1 kilogram of air 1 degree would have for its value 0.267. Thus the quantity of heat furnished by the body *A* is

0.267 units.

This is the heat capable of producing 0.000000372 units of motive power by its fall from $0^{\circ}.001$ to zero.

For a fall a thousand times greater, for a fall of one degree, the motive power will be very nearly a thousand times the former, or

$$0.000372.$$

If, now, instead of 0.267 units of heat we employ 1000 units, the motive power produced will be expressed by the proportion

$$\frac{0.267}{0.000372} = \frac{1000}{x}, \text{ whence } x = \frac{372}{267} = 1.395.$$

Thus 1000 units of heat passing from a body maintained at the temperature of 1 degree to another body maintained at zero would produce, in acting upon the air,

1.395 units of motive power.*

We will now compare this result with that furnished by the action of heat on the vapor of water.

Let us suppose one kilogram of liquid water enclosed in the cylindrical vessel *abcd* (Fig. 4), between the bottom *ab* and the piston *cd*.

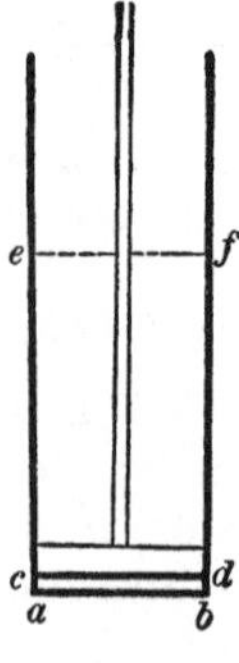

Fig. 4

Let us suppose, also, the two bodies *A*, *B* maintained each at a constant temperature, that of *A* being a very little above that of *B*. Let us imagine now the following operations:

(1) Contact of the water with the body *A*, movement of the piston from the position *cd* to the position *ef*, formation of steam at the

* [The correct result should be 1.56 units since $J=0.427$ of Carnot's units of work per kilocalorie. The discrepancy is due to a 15% inconsistency between Poisson's datum and the measurements of Delaroche and Bérard. E. M.]

temperature of the body A to fill the vacuum produced by the extension of volume. We will suppose the space *abef* large enough to contain all the water in a state of vapor.

(2) Removal of the body A, contact of the vapor with the body B, precipitation of a part of this vapor, diminution of its elastic force, return of the piston from *ef* to *ab*, liquefaction of the rest of the vapor through the effect of the pressure combined with the contact of the body B.

(3) Removal of the body B, fresh contact of the water with the body A, return of the water to the temperature of this body, renewal of the former period, and so on.

The quantity of motive power developed in a complete cycle of operations is measured by the product of the volume of the vapor multiplied by the difference between the tensions that it possesses at the temperature of the body A and at that of the body B. As to the heat employed, that is to say, transported from the body A to the body B, it is evidently that which was necessary to turn the water into vapor, disregarding always the small quantity required to restore the temperature of the liquid water from that of B to that of A.

Suppose the temperature of the body A 100 degrees, and that of the body B 99 degrees: the difference of the tensions will be, according to the table of M. Dalton, 26 millimeters of mercury or $0^{m}.36$ head of water.

The volume of the vapor is 1700 times that of the water. If we operate on one kilogram, that will be 1700 liters, or $1^{mc}.700$.

Thus the value of the motive power developed is the product

$$1.700 \times 0.36 = 0.611 \text{ units},$$

of the kind of which we have previously made use.

The quantity of heat employed is the quantity required to turn into vapor water already heated to 100°. This quantity is found by experiment. We have found it equal to 550°, or, to speak more exactly, to 550 of our units of heat.

Thus 0.611 units of motive power result from the employment of 550 units of heat. The quantity of motive power resulting from 1000 units of heat will be given by the proportion

$$\frac{550}{0.611} = \frac{1000}{x}, \text{ whence } x = \frac{611}{550} = 1.112.$$

Thus 1000 units of heat transported from one body kept at 100 degrees to another kept at 99 degrees will produce, acting upon vapor of water, 1.112 units of motive power.*

The number 1.112 differs by about $\frac{1}{4}$ from the number 1.395 previously found for the value of the motive power developed by 1000 units of heat acting upon the air; but it should be observed that in this case the temperatures of the bodies *A* and *B* were 1 degree and zero, while here they are 100 degrees and 99 degrees. The difference is much the same; but it is not found at the same height in the thermometric scale. To make an exact comparison, it would have been necessary to estimate the motive power developed by the steam formed at 1 degree and condensed at zero. It would also have been necessary to know the quantity of heat contained in the steam formed at 1 degree.

The law of MM. Clément and Désormes referred to on page 33 gives us this datum. The constituent heat of vapor of water being always the same at any temperature at which vaporization takes place, if 550 degrees of heat are required to vaporize water already brought up to 100 degrees, 550+100 or 650 will be required to vaporize the same weight of water taken at zero.

Making use of this datum and reasoning exactly as we did for water at 100 degrees, we find, as is easily seen,

$$1.290$$

for the motive power developed by 1000 units of heat acting upon the vapor of water between one degree and zero. This number approximates more closely than the first to

$$1.395.$$

It differs from it only $\frac{1}{13}$, an error which does not exceed probable limits, considering the great number of data of different sorts of which we have been obliged to make use in order to arrive at this approximation. Thus is our fundamental law verified in a special case.†

---

* [This uses the Clapeyron relation between latent heat and the slope of the vapor-pressure curve. The numerical result at 100°C is correct to 2%, but Watt's law and Dalton's data give a 20% error at 0°C. E. M.]

† We find (*Annales de Chimie et de Physique*, July, 1818, p. 294) in a memoir of M. Petit an estimate of the motive power of heat applied to air and to vapor of water. This estimate leads us to attribute a great advantage to atmospheric air, but it is derived by a method of considering the action of heat which is quite imperfect.

We will examine another case in which vapor of alcohol is acted upon by heat. The reasoning is precisely the same as for the vapor of water. The data alone are changed. Pure alcohol boils under ordinary pressure at 78°.7 Centigrade. One kilogram absorbs, according to MM. Delaroche and Bérard, 207 units of heat in undergoing transformation into vapor at this same temperature, 78°.7.

The tension of the vapor of alcohol at one degree below the boiling-point is found to be diminished $\frac{1}{25}$. It is $\frac{1}{25}$ less than the atmospheric pressure; at least, this is the result of the experiment of M. Bétancour reported in the second part of *l'Architecture hydraulique* of M. Prony, pp. 180, 195.*

If we use these data, we find that, in acting upon one kilogram of alcohol at the temperatures of 78°.7 and 77°.7, the motive power developed will be 0.251 units.

This results from the employment of 207 units of heat. For 1000 units the proportion must be

$$\frac{207}{0.254} = \frac{1000}{x}, \text{ whence } x = 1.230.\dagger$$

This number is a little more than the 1.112 resulting from the use of the vapor of water at the temperatures 100° and 99°; but if we suppose the vapor of water used at the temperatures 78° and 77°, we find, according to the law of MM. Clément and Désormes,

* M. Dalton believed that he had discovered that the vapors of different liquids at equal thermometric distances from the boiling-point possess equal tensions; but this law is not precisely exact; it is only approximate. It is the same with the law of the proportionality of the latent heat of vapors with their densities (see Extracts from a Memoir of M. C. Despretz, *Annales de Chimie et de Physique*, xvi, p. 105, and xxiv, p. 323). Questions of this nature are closely connected with those of the motive power of heat. Quite recently MM. H. Davy and Faraday, after having conducted a series of elegant experiments on the liquefaction of gases by means of considerable pressure, have tried to observe the changes of tension of these liquefied gases on account of slight changes of temperature. They have in view the application of the new liquids to the production of motive power (see *Annales de Chimie et de Physique*, January, 1824, p. 80).

According to the above-mentioned theory, we can foresee that the use of these liquids would present no advantages relatively to the economy of heat. The advantages would be found only in the lower temperature at which it would be possible to work, and in the sources whence, for this reason, it would become possible to obtain caloric.

† [These results are correct to 2%, Watt's law being quite accurate at 78°C. E. M.]

1.212 for the motive power due to 1000 units of heat. This latter number approaches, as we see, very nearly to 1.230. There is a difference of only $\frac{1}{50}$.

We should have liked to be able to make other approximations of this sort—to be able to calculate, for example, the motive power developed by the action of heat on solids and liquids, by the congelation of water, and so on; but physics as yet refuses us the necessary data.*

The fundamental law that we proposed to confirm seems to us to require, however, in order to be placed beyond doubt, new verifications. It is based upon the theory of heat as it is understood today, and it should be said that this foundation does not appear to be of unquestionable solidity. New experiments alone can decide the question. Meanwhile we can apply the theoretical ideas expressed above, regarding them as exact, to the examination of the different methods proposed up to date, for the realization of the motive power of heat.†

It has sometimes been proposed to develop motive power by the action of heat on solid bodies. The mode of procedure which naturally first occurs to the mind is to fasten immovably a solid body—a metallic bar, for example—by one of its extremities; to attach the other extremity to a movable part of the machine; then, by successive heating and cooling, to cause the length of the bar to vary, and so to produce motion. Let us try to decide whether this method of developing motive power can be advantageous. We have shown that the condition of the most effective employment of heat in the production of motion is, that all changes of temperature occurring in the bodies should be due to changes of volume. The nearer we come to fulfilling this condition the more fully will the

---

* Those that we need are the expansive force acquired by solids and liquids by a given increase of temperature, and the quantity of heat absorbed or relinquished in the changes of volume of these bodies.

† [In the *manuscript* of the *Reflections*, this paragraph is entirely different in tone: "The fundamental law that we proposed to confirm seems to us to have been placed beyond doubt, both by the reasoning which served to establish it, and by the calculations which have just been made. We will now apply the theoretical ideas expressed above to the examination of the different methods proposed to date for the realization of the motive power of heat." It is certain that by the time the book was finished, Carnot had already begun to doubt the basic fact that the quantity of heat in a body was uniquely determined by the pressure and temperature. The whole idea of the cycle of operations was therefore suspect, and Carnot presumably altered the paragraph in proof. E. M.]

heat be utilized. Now, working in the manner just described, we are very far from fulfilling this condition: change of temperature is not due here to change of volume; all the changes are due to contact of bodies differently heated—to the contact of the metallic bar, either with the body charged with furnishing heat to it, or with the body charged with carrying it off.

The only means of fulfilling the prescribed condition would be to act upon the solid body exactly as we did on the air in the operations described on page 17. But for this we must be able to produce, by a change only of volume of the solid body, considerable changes of temperature, that is, if we should want to utilize considerable falls of caloric. Now this appears impracticable. In short, many considerations lead to the conclusion that the changes produced in the temperature of solid or liquid bodies through the effect of compression and rarefaction would be but slight.

(1) We often observe in machines (particularly in steam-engines) solid pieces which endure considerable strain in one way or another, and although these efforts may be sometimes as great as the nature of the substances employed permits, the variations of temperature are scarcely perceptible.

(2) In the action of striking medals, in that of the rolling-mill, of the draw plate, the metals undergo the greatest compression to which we can submit them, employing the hardest and strongest tools. Nevertheless the elevation of temperature is not great. If it were, the pieces of steel used in these operations would soon lose their temper.

(3) We know that it would be necessary to exert on solids and liquids a very great strain in order to produce in them a reduction of volume comparable to that which they experience in cooling (cooling from 100° to zero, for example). Now the cooling requires a greater abstraction of caloric than would simple reduction of volume. If this reduction were produced by mechanical means, the heat set free would not then be able to make the temperature of the body vary as many degrees as the cooling makes it vary. It would, however, necessitate the employment of a force undoubtedly very considerable.

Since solid bodies are susceptible of little change of temperature through changes of volume, and since the condition of the most effective employment of heat for the development of motive power is precisely that all change of temperature should be due to a change of volume, solid bodies appear but ill fitted to realize this power.

The same remarks apply to liquids. The same reasons may be given for rejecting them.*

We are not speaking now of practical difficulties. They will be numberless. The motion produced by the dilatation and compression of solid or liquid bodies would only be very slight. In order to give them sufficient amplitude we should be forced to make use of complicated mechanisms. It would be necessary to employ materials of the greatest strength to transmit enormous pressure; finally, the successive operations would be executed very slowly compared to those of the ordinary steam-engine, so that apparatus of large dimensions and heavy cost would produce but very ordinary results.

The elastic fluids, gases or vapors, are the means really adapted to the development of the motive power of heat. They combine all the conditions necessary to fulfil this office. They are easy to compress; they can be almost infinitely expanded; variations of volume occasion in them great changes of temperature; and, lastly, they are very mobile, easy to heat and to cool, easy to transport from one place to another, which enables them to produce rapidly the desired effects. We can easily conceive a multitude of machines fitted to develop the motive power of heat through the use of elastic fluids; but in whatever way we look at it, we should not lose sight of the following principles:

(1) The temperature of the fluid should be made as high as possible, in order to obtain a great fall of caloric, and consequently a large production of motive power.

(2) For the same reason the cooling should be carried as far as possible.

(3) It should be so arranged that the passage of the elastic fluid from the highest to the lowest temperature should be due to increase of volume; that is, it should be so arranged that the cooling of the gas should occur spontaneously as the effect of rarefaction.

The limits of the temperature to which it is possible to bring the fluid primarily, are simply the limits of the temperature obtainable by combustion; they are very high.

The limits of cooling are found in the temperature of the coldest body of which we can easily and freely make use; this body is usually the water of the locality.

---

* The recent experiments of M. Oerstedt on the compressibility of water have shown that, for a pressure of five atmospheres, the temperature of this liquid exhibits no appreciable change. (See *Annales de Chimie et de Physique*, Feb. 1823, p. 192.)

As to the third condition, it involves difficulties in the realization of the motive power of heat when the attempt is made to take advantage of great differences of temperature, to utilize great falls of heat. In short, it is necessary then that the gas, by reason of its rarefaction, should pass from a very high temperature to a very low one, which requires a great change of volume and of density, which requires also that the gas be first taken under a very heavy pressure, or that it acquire by its dilatation an enormous volume—conditions both difficult to fulfill. The first necesitates the employment of very strong vessels to contain the gas at a very high temperature and under very heavy pressure. The second necessitates the use of vessels of large dimensions. These are, in a word, the principal obstacles which prevent the utilization in steam-engines of a great part of the motive power of the heat. We are obliged to limit ourselves to the use of a slight fall of caloric, while the combustion of the coal furnishes the means of procuring a very great one.

It is seldom that in steam-engines the elastic fluid is produced under a higher pressure than six atmospheres, a pressure corresponding to about 160° Centigrade, and it is seldom that condensation takes place at a temperature much under 40°. The fall of caloric from 160° to 40° is 120°, while by combustion we can procure a fall of 1000° to 2000°.

In order to comprehend this more clearly, let us recall what we have termed the fall of caloric. This is the passage of the heat from one body, *A*, having an elevated temperature, to another, *B*, where it is lower. We say that fall of the caloric is 100° or 1000° when the difference of temperature between the bodies *A* and *B* is 100° or 1000°.

In a steam-engine which works under a pressure of six atmospheres the temperature of the boiler is 160°. This is the body *A*. It is kept, by contact with the furnace, at the constant temperature of 160°, and continually furnishes the heat necessary for the formation of steam. The condenser is the body *B*. By means of a current of cold water it is kept at a nearly constant temperature of 40°. It absorbs continually the caloric brought from the body *A* by the steam. The difference of temperature between these two bodies is 160° – 40°, or 120°. Hence we say that the fall of caloric is here 120°.

Coal being capable of producing, by its combustion, a temperature higher than 1000°, and the cold water, which is generally used in our climate, being at about 10°, we can easily procure a fall of caloric of 1000°, and of this only 120° are utilized by steam-engines.

Even these 120° are not wholly utilized. There is always considerable loss due to useless re-establishments of equilibrium in the caloric.

It is easy to see the advantages possessed by high-pressure machines over those of lower pressure. *This superiority lies essentially in the power of utilizing a greater fall of caloric.* The steam produced under a higher pressure is found also at a higher temperature, and as, further, the temperature of condensation remains always about the same, it is evident that the fall of caloric is more considerable. But to obtain from high-pressure engines really advantageous results, it is necessary that the fall of caloric should be most profitably utilized. It is not enough that the steam be produced at a high temperature: it is also necessary that by the expansion of its volume its temperature should become sufficiently low. A good steam-engine, therefore, should not only employ steam under heavy pressure, but *under successive and very variable pressures, differing greatly from one another, and progressively decreasing.**

---

* This principle, the real foundation of the theory of steam-engines, was very clearly developed by M. Clément in a memoir presented to the Academy of Sciences several years ago. This memoir has never been printed, and I owe the knowledge of it to the kindness of the author. Not only is the principle established therein, but it is applied to the different systems of steam-engines actually in use. The motive power of each of them is estimated therein by the aid of the law cited, page 33, and compared with the results of experiment.

The principle in question is so little known or so poorly appreciated, that recently Mr. Perkins, a celebrated mechanician of London, constructed a machine in which steam produced under the pressure of 35 atmospheres—a pressure never before used—is subjected to very little expansion of volume, as any one with the least knowledge of this machine can understand. It consists of a single cylinder of very small dimensions, which at each stroke is entirely filled with steam, formed under the pressure of 35 atmospheres. The steam produces no effect by the expansion of its volume, for no space is provided in which the expansion can take place. It is condensed as soon as it leaves the small cylinder. It works therefore only under a pressure of 35 atmospheres, and not, as its useful employment would require, under progressively decreasing pressures. The machine of Mr. Perkins seems not to realize the hopes which it at first awakened. It has been asserted that the economy of coal in this engine was $\frac{9}{10}$ above the best engines of Watt, and that it possessed still other advantages (see *Annales de Chimie et de Physique*, April, 1823, p. 429). These assertions have not been verified. The engine of Mr. Perkins is nevertheless a valuable invention, in that it has proved the possibility of making use of steam under much higher pressure than previously, and because, being easily modified, it may lead to very useful results.

Watt, to whom we owe almost all the great improvements in steam-engines, and who brought these engines to a state of perfection difficult even now to surpass, was also the first who employed steam under progressively decreasing pressures. In many cases he suppressed the introduction of the steam into the cylinder at a half, a third, or a quarter of the stroke. The piston completes its stroke, therefore, under a constantly diminishing pressure. The first engines working on this principle date from 1778. Watt conceived the idea of them in 1769, and took out a patent in 1782.

We give here the Table appended to Watt's patent. He supposed the steam introduced into the cylinder during the first quarter of the stroke of the piston; then, dividing this stroke into twenty parts, he calculated the mean pressure as follows:

Portions of the descent from the top of the cylinder			Decreasing pressure of the steam, the full pressure being 1	
	0.05	Steam arriving freely from the boiler	1.000	Full pressure
	0.10		1.000	
	0.15		1.000	
	0.20		1.000	
Quarter	0.25		1.000	
	0.30	The steam being cut off and the descent taking place only by expansion	0.830	
	0.35		0.714	
	0.40		0.625	
	0.45		0.555	Half original pressure
Half	0.50		0.500	
	0.55		0.454	
	0.60		0.417	
	0.65		0.385	
	0.70		0.375	
	0.75		0.333	One third
	0.80		0.312	
	0.85		0.294	
	0.90		0.277	
	0.95		0.262	
Bottom of cylinder	1.00		0.250	Quarter
		Total,	11.583	

$$\text{Mean pressure } \frac{11.583}{20} = 0.579.$$

On which he remarked, that the mean pressure is more than half the original pressure; also that in employing a quantity of steam equal to a quarter, it would produce an effect more than half.

Watt here supposed that steam follows in its expansion the law of Mariotte, which should not be considered exact, because, in the first place, the elastic fluid in dilating falls in temperature, and in the second place there is nothing to prove that a part of this fluid is not condensed by its expansion. Watt should also have taken into consideration the force necessary to expel the steam which remains after condensation, and which is found in quantity as much greater as the expansion of the volume has been carried further. Dr. Robison has supplemented the work of Watt by a simple formula to calculate the effect of the expansion of steam, but this formula is found to have the same faults that we have just noticed. It has nevertheless been useful to constructors by furnishing them approximate data practically quite satisfactory. We have considered it useful to recall these facts because they are little known, especially in France. These engines have been built after the models of the inventors, but the ideas by which the inventors were

In order to understand in some sort *à posteriori* the advantages of high-pressure engines, let us suppose steam to be formed under atmospheric pressure and introduced into the cylindrical vessel *abcd* (Fig. 5), under the piston *cd*, which at first touches the bottom *ab*. The steam, after having moved the piston from *ab* to *cd*, will

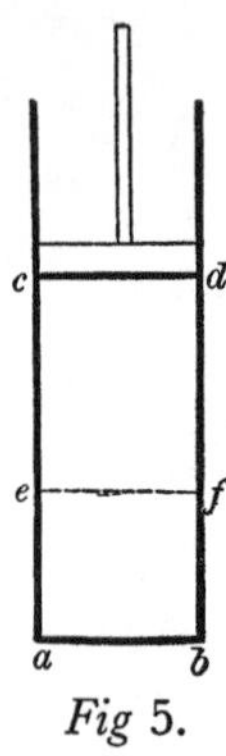

Fig 5.

continue finally to produce its results in a manner with which we will not concern ourselves.

Let us suppose that the piston having moved to *cd* is forced downward to *ef*, without the steam being allowed to escape, or any portion of its caloric to be lost. It will be driven back into the space *abef*, and will increase at the same time in density, elastic force, and temperature. If the steam, instead of being produced under atmospheric pressure, had been produced just when it was being forced back into *abef*, and so after its introduction into the cylinder it had made the piston move from *ab* to *ef*, and had moved it simply by its extension of volume, from *ef* to *cd*, the motive power produced would have been more considerable than in the first case. In fact, the movement of the piston, while equal in extent, would have taken place under the action of a greater pressure, though variable, and though progressively decreasing.

The steam, however, would have required for its formation

---

originally influenced have been but little understood. Ignorance of these ideas has often led to grave errors. Engines originally well conceived have deteriorated in the hands of unskillful builders, who, wishing to introduce in them improvements of little value, have neglected the capital considerations which they did not know enough to appreciate.

exactly the same quantity of caloric, only the caloric would have been employed at a higher temperature.

It is considerations of this nature which have led to the making of double-cylinder engines—engines invented by Mr. Hornblower, improved by Mr. Woolf, and which, as regards economy of the combustible, are considered the best. They consist of a small cylinder, which at each pulsation is filled more or less (often entirely) with steam, and of a second cylinder having usually a capacity quadruple that of the first, and which receives no steam except that which has already operated in the first cylinder. Thus the steam when it ceases to act has at least quadrupled in volume. From the second cylinder it is carried directly into the condenser, but it is conceivable that it might be carried into a third cylinder quadruple the second, and in which its volume would have become sixteen times the original volume. The principal obstacle to the use of a third cylinder of this sort is the capacity which it would be necessary to give it, and the large dimensions which the openings for the passage of the steam must have.* We will say no more on this subject, as we do not propose here to enter into the details of construction of steam-engines. These details call for a work devoted specially to them, and which does not yet exist, at least in France.†

* The advantage in substituting two cylinders for one is evident. In a single cylinder the impulsion of the piston would be extremely variable from the beginning to the end of the stroke. It would be necessary for all the parts by which the motion is transmitted to be of sufficient strength to resist the first impulsion, and perfectly fitted to avoid the abrupt movements which would greatly injure and soon destroy them. It would be especially on the working beam, on the supports, on the crank, on the connecting rod, and on the first gear wheels that the unequal effort would be felt, and would produce the most injurious effects. It would be necessary that the steam cylinder should be both sufficiently strong to sustain the highest pressure, and with a large enough capacity to contain the steam after its expansion of volume, while in using two successive cylinders it is only necessary to have the first sufficiently strong and of medium capacity—which is not at all difficult—and to have the second of ample dimensions, with moderate strength.

Double-cylinder engines, although founded on correct principles, often fail to secure the advantages expected from them. This is due principally to the fact that the dimensions of the different parts of these engines are difficult to adjust, and that they are rarely found to be in correct proportion. Good models for the construction of double-cylinder engines are wanting, while excellent designs exist for the construction of engines on the plan of Watt. From this arises the diversity that we see in the results of the former, and the great uniformity that we have observed in the results of the latter.

† We find in the work called *De la Richesse Minérale*, by M. Héron de Villefosse, vol. iii, p. 50 and following, a good description of the steam-engines actually in

If the expansion of the steam is mainly limited by the dimensions of the vessels in which the dilatation must take place, the degree of condensation at which it is possible to use it at first is limited only by the resistance of the vessels in which it is produced, that is, of the boilers.

In this respect we have by no means attained the best possible results. The arrangement of the boilers generally in use is entirely faulty, although the tension of the steam rarely exceeds from four to six atmospheres. They often burst and cause severe accidents. It will undoubtedly be possible to avoid such accidents, and meantime to raise the steam to much greater pressures than is usually done.

Besides the high-pressure double-cylinder engines of which we have spoken, there are also high-pressure engines of one cylinder. The greater part of these latter have been constructed by two ingenious English engineers, Messrs. Trevithick and Vivian. They employ the steam under a very high pressure, sometimes eight to ten atmospheres, but they have no condenser. The steam, after it has been introduced into the cylinder, undergoes therein a certain increase of volume, but preserves always a pressure higher than atmospheric. When it has fulfilled its office it is thrown out into the atmosphere. It is evident that this mode of working is fully equivalent, in respect to the motive power produced, to condensing the steam at 100°, and that a portion of the useful effect is lost. But the engines working thus dispense with condenser and air pump. They are less costly than the others, less complicated, occupy less space, and can be used in places where there is not sufficient water for condensation. In such places they are of inestimable advantage, since no others could take their place. These engines are principally employed in England to move coal wagons on railroads laid either in the interior of mines or outside them.

We have, further, only a few remarks to make upon the use of permanent gases and other vapors than that of water in the development of the motive power of heat.

Various attempts have been made to produce motive power by the action of heat on atmospheric air. This gas presents, as compared with vapor of water, both advantages and disadvantages, which we will proceed to examine.

---

use in mining. In England the steam-engine has been very fully discussed in the *Encyclopedia Britannica*. Some of the data here employed are drawn from the latter work.

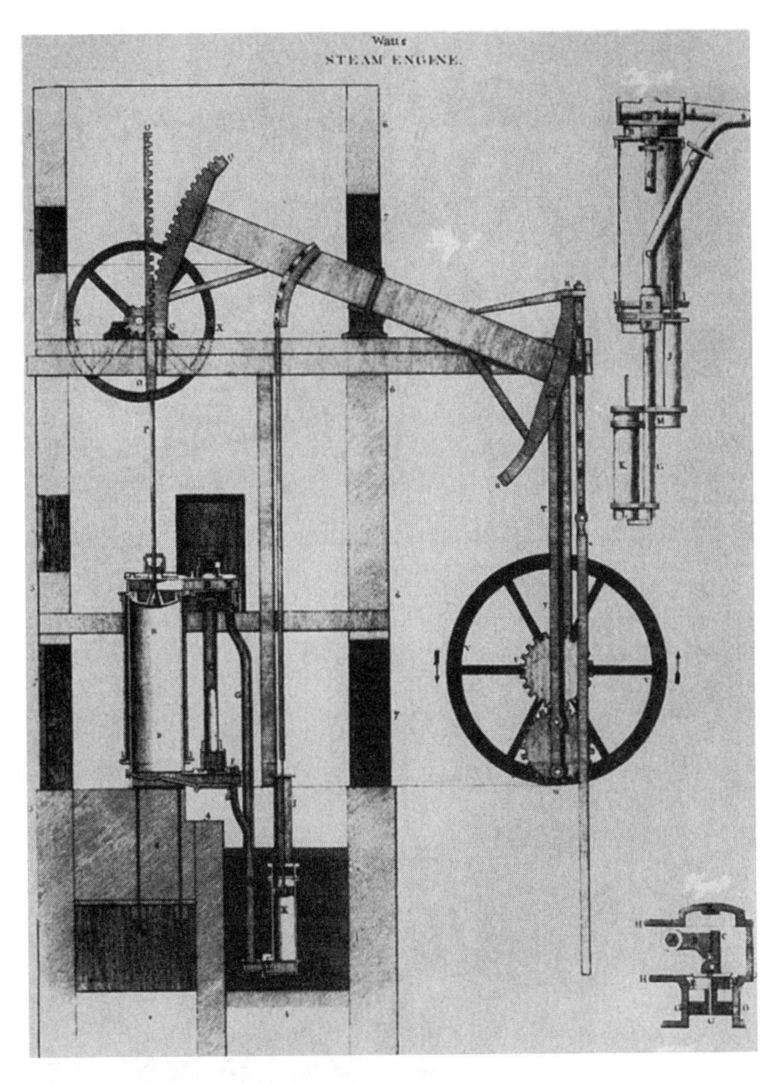

The next eight illustrations are of engines discussed by Carnot in the Reflections. *This is an early design by James Watt. The steam was generated at atmospheric pressure and acted through the rarefaction produced by its condensation in the small cylinder immersed in cold water. This machine was the first to be double-acting—that is, to work on both the up and down strokes—and to use steam expansively; it was one of the first to use the sun-and planet motion to drive a wheel.* (*From the* Encyclopaedia Britannica, *1797, based on a Patent Specification.*)

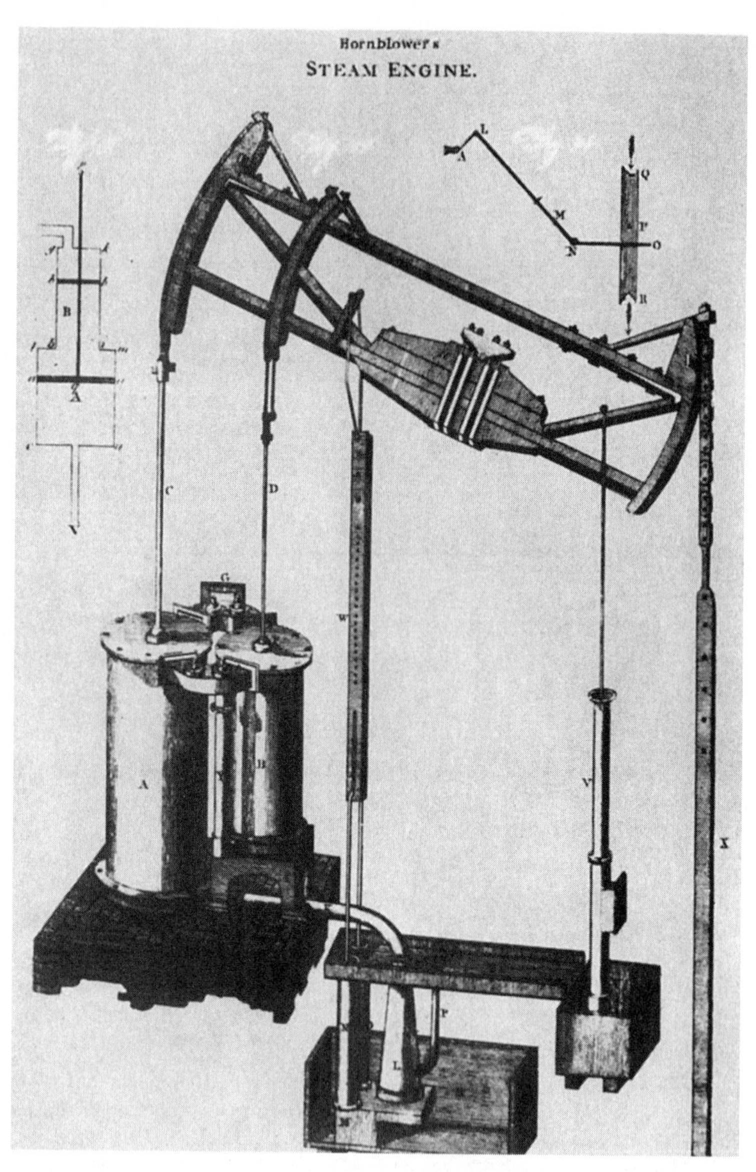

This is an early two-cylinder or compound engine by Jonathan Hornblower of Bristol. The steam first entered the small high-pressure cylinder and then the larger low-pressure one. (From the Encyclopaedia Britannica, *1810.)*

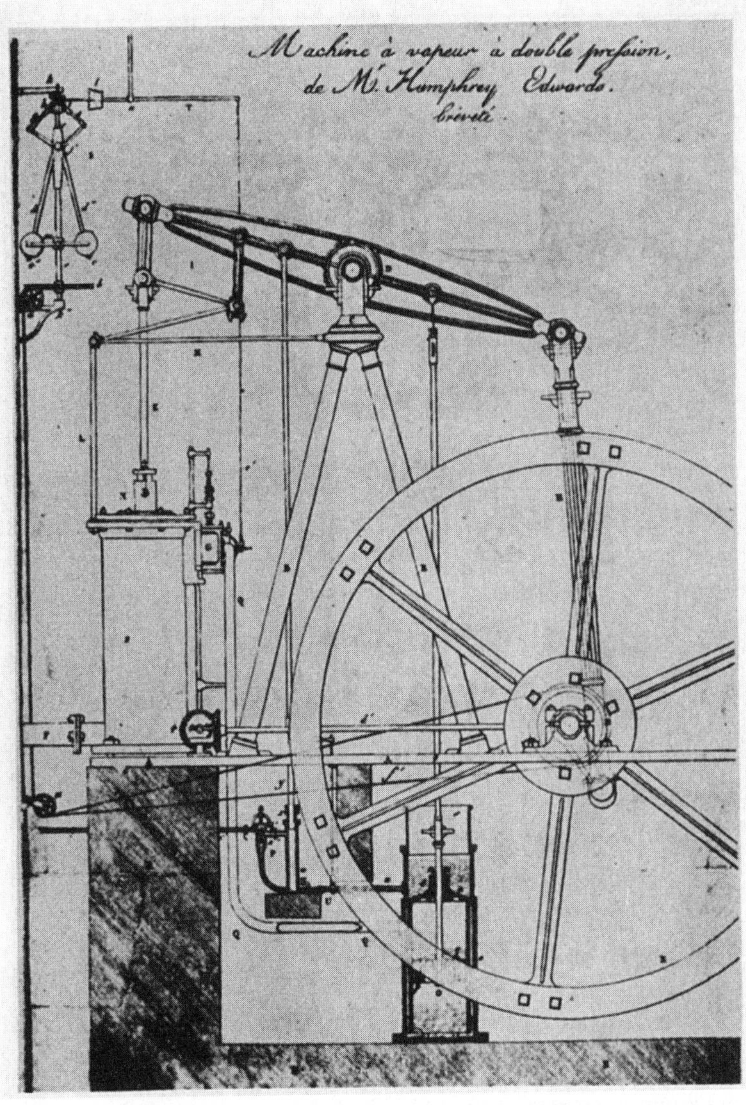

Arthur Woolf made a commercial success of the compound engine in spite of opposition from James Watt. This small one was erected in 1815 in Paris by Edwards and Company. The two cylinders were inside the same steam chest. Woolf's boilers were of bad design and silted up, so that the efficiency of his machines always decreased with time. (*From* Transactions of the Newcomen Society, *XIII, 55, 1934.*)

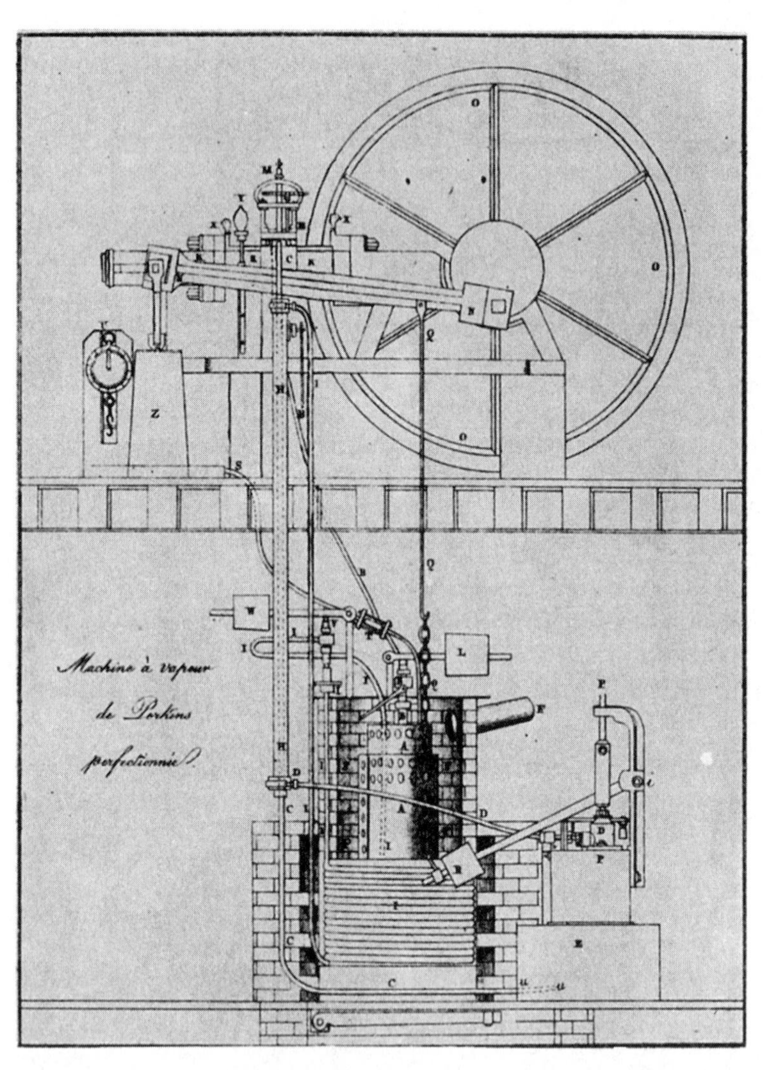

Jacob Perkins' high-pressure steam-engine was claimed to have ten times the efficiency of ordinary machines. The 1-inch diameter cylinder was inside the arm K at the top left. A heat exchanger H carrying waste steam at high pressure ("surcharged with caloric") was claimed to heat the counter-flowing low-pressure water to a temperature hotter than the steam. Perkins was born in Massachusetts, came to London and printed the "penny blacks," the world's first postage stamps. (From the Bibliothèque Universelle, *XXV, 1824.)*

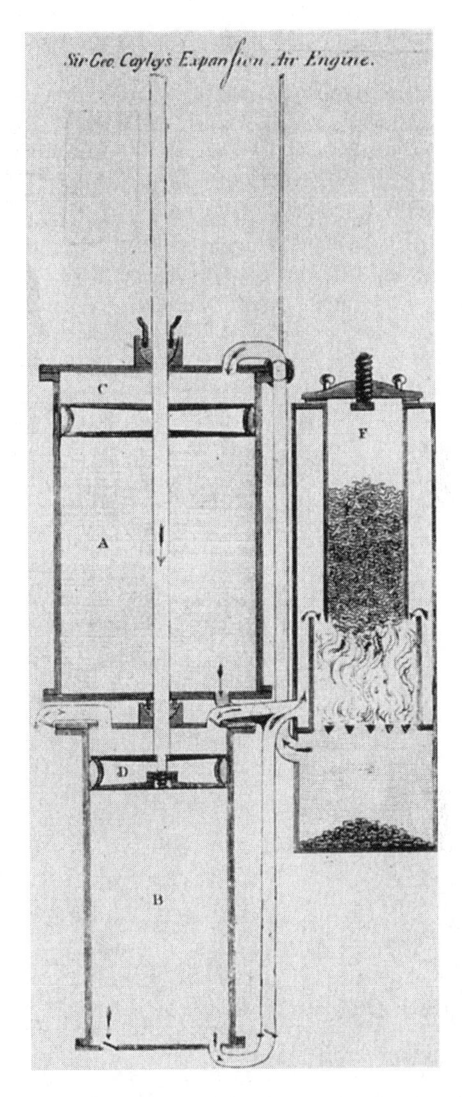

Sir George Cayley's air-engine, proposed as "a light locomotive engine for travelling on turnpike roads." The lower cylinder was a double-acting compressor driving air through the fire into the upper cylinder. It did not work very well, he said, because the cylinders were made from copper sheet. Cayley, "the father of aeronautics," made the first gliders and constructed a gunpowder engine in 1807 to propel one by rowing with oars. (From Nicholson's Journal, *XVIII, 260, 1807.)*

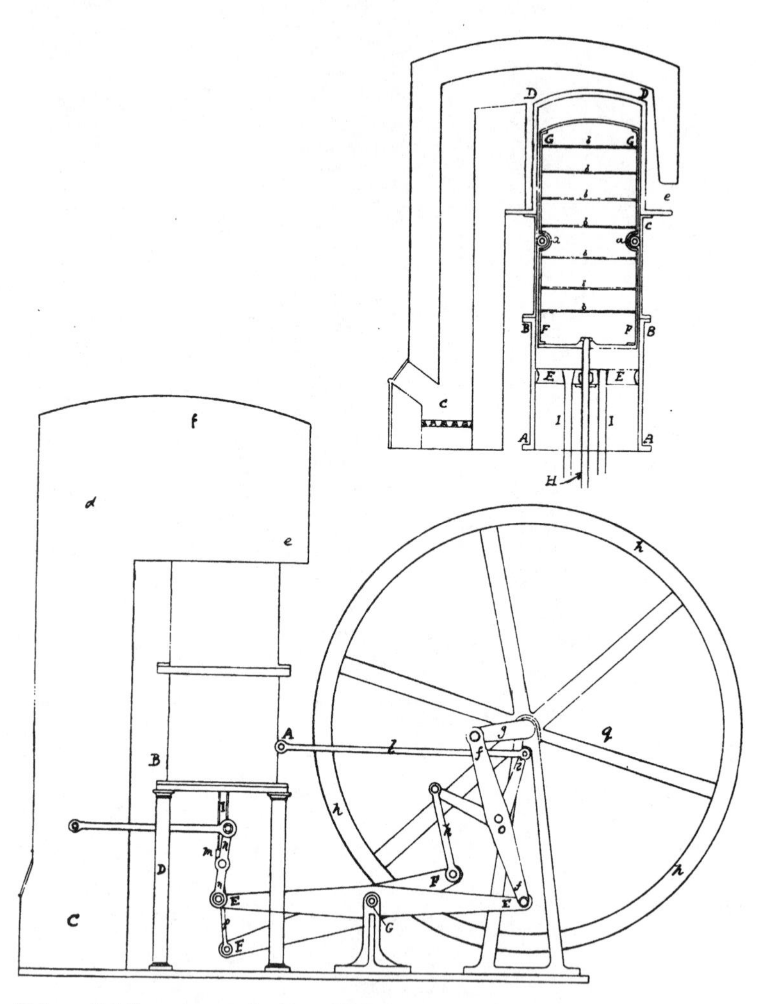

Robert Stirling patented this air-engine in 1816. Air was compressed and expanded in a closed cycle by a displacer with regenerator. The cycle was thermodynamically as efficient as the Carnot cycle and is now the basis for liquid air refrigeration. But as late as 1845 Stirling stated that "the economy depended on the reiterated use of the same air alternately giving out and absorbing the same caloric. . . ." (From The Engineer, *CXXIV, 516, 1917.)*

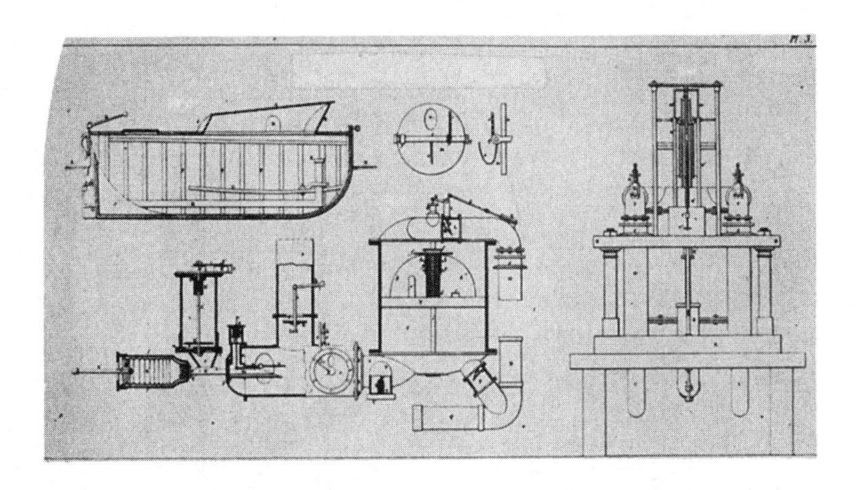

The pyréolophore (1806), surely one of the most fascinating engines ever invented. It was an internal-combustion machine burning lycopodium powder, ignited by a gunner's match. Lazare Carnot and Berthollet reported that it made 13 strokes a minute and drove a half-ton boat against the current of the river Saône. Other versions could work off powdered coal mixed with resin, and petroleum oil. The Niepce cousins also invented an early photographic process and the hobby horse, ancestor of the bicycle. (From the Musée Denon, Chalon-sur-Saône.)

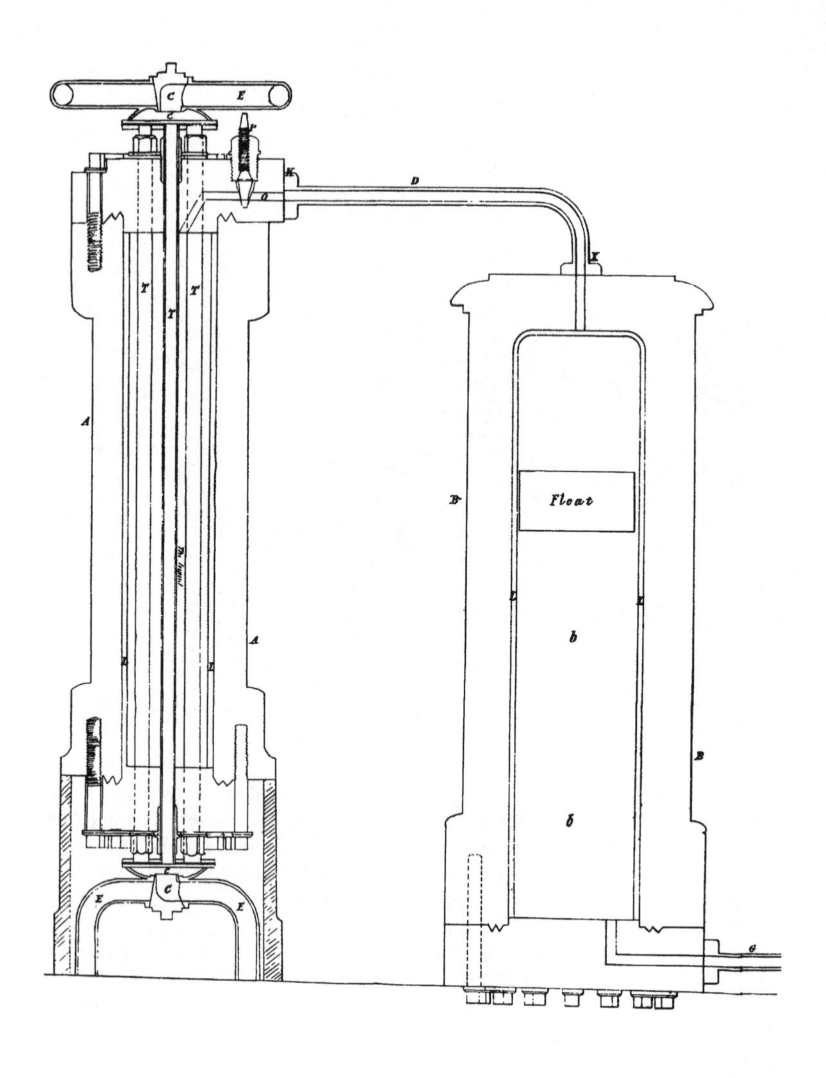

News of this engine (patented by Marc Isambard Brunel) using liquid carbon dioxide reached Paris not long after Carnot's book was published. The liquid gas was to be forced into the wood-lined cyclinder on the left, which was fitted with tubes carrying hot and cold water alternately. The gas pressure, estimated to alternate between 40 and 90 atmospheres, was to be transmitted through oil in the vessel on the right to a piston (not shown). (From a Patent Specification.)

(1) It presents, as compared with vapor of water, a notable advantage in that, having for equal volume a much less capacity for heat, it would cool more rapidly by an equal increase of volume. (This fact is proved by what has already been stated.) Now we have seen how important it is to produce by change of volume the greatest possible changes of temperature.

(2) Vapors of water can be formed only through the intervention of a boiler, while atmospheric air could be heated directly by combustion carried on within its own mass. Considerable loss could thus be prevented, not only in the quantity of heat, but also in its temperature. This advantage belongs exclusively to atmospheric air. Other gases do not possess it. They would be even more difficult to heat than vapor of water.

(3) In order to give to air great increase of volume, and by that expansion to produce a great change of temperature, it must first be taken under a sufficiently high pressure; then it must be compressed with a pump or by some other means before heating it. This operation would require a special apparatus, an apparatus not found in steam-engines. In the latter, water is in a liquid state when injected into the boiler, and to introduce it requires but a small pump.

(4) The condensing of the vapor by contact with the refrigerant body is much more prompt and much easier than is the cooling of air. There might, of course, be the expedient of throwing the latter out into the atmosphere, and there would be also the advantage of avoiding the use of a refrigerant, which is not always available, but it would be requisite that the increase of the volume of the air should not reduce its pressure below that of the atmosphere.

(5) One of the gravest inconveniences of steam is that it cannot be used at high temperatures without necessitating the use of vessels of extraordinary strength. It is not so with air for which there exists no necessary relation between the elastic force and the temperature. Air, then, would seem more suitable than steam to realize the motive power of falls of caloric from high temperatures; perhaps at low temperatures steam may be more convenient. We might conceive even the possibility of making the same heat act successively upon air and vapor of water. It would be only necessary that the air should have, after its use, an elevated temperature, and instead of throwing it out immediately into the atmosphere, to make it envelop a steam boiler, as if it issued directly from a furnace.

The use of atmospheric air for the development of the motive power of heat presents in practice very great, but perhaps not insurmountable, difficulties. If we should succeed in overcoming them, it would doubtless offer a notable advantage over vapor of water.*

As to the other permanent gases, they should be absolutely rejected. They have all the inconveniences of atmospheric air, with none of its advantages. The same can be said of other vapors than that of water, as compared with the latter.

If we could find an abundant liquid body which would vaporize at a higher temperature than water, of which the vapor would have,

---

* Among the attempts made to develop the motive power of heat by means of atmospheric air, we should mention those of MM. Niepce, which were made in France several years ago, by means of an apparatus called by the inventors a pyréolophore. The apparatus was made thus: There was a cylinder furnished with a piston, into which the atmospheric air was introduced at ordinary density. A very combustible material, reduced to a condition of extreme tenuity, was thrown into it, remained a moment in suspension in the air, and then flame was applied. The inflammation produced very nearly the same effect as if the elastic fluid had been a mixture of air and combustible gas, of air and carburetted hydrogen gas, for example. There was a sort of explosion, and a sudden dilatation of the elastic fluid—a dilatation that was utilized by making it act upon the piston. The latter may have a motion of any amplitude whatever, and the motive power is thus realized. The air is next renewed, and the operation repeated.

This machine, very ingenious and interesting, especially on account of the novelty of its principle, fails in an essential point. The material used as a combustible (it was the dust of Lycopodium, used to produce flame in our theatres) was so expensive, that all the advantage was lost through that cause; and unfortunately it was difficult to employ a combustible of moderate price, since a very finely powdered substance was required which would burn quickly, spread rapidly, and leave little or no ash.

Instead of working as did MM. Niepce, it would seem to us preferable to compress the air by means of pumps, to make it traverse a perfectly closed furnace into which the combustible had been introduced in small portions by a mechanism easy of conception, to make it develop its action in a cylinder with a piston, or in any other variable space; finally, to throw it out again into the atmosphere, or even to make it pass under a steam boiler in order to utilize the temperature remaining.

The principal difficulties that we should meet in this mode of operation would be to enclose the furnace in a sufficiently strong envelope, to keep the combustion meanwhile in the requisite state, to maintain the different parts of the apparatus at a moderate temperature, and to prevent rapid abrasion of the cylinder and of the piston. These difficulties do not appear to be insurmountable.

There have been made, it is said, recently in England, successful attempts to develop motive power through the action of heat on atmospheric air. We are entirely ignorant in what these attempts have consisted—if indeed they have really been made.

for the same volume, a less specific heat, which would not attack the metals employed in the construction of machines, it would undoubtedly merit the preference. But nature provides no such body.

The use of the vapor of alcohol has been proposed. Machines have even been constructed for the purpose of using it, by avoiding the mixture of its vapor with the water of condensation, that is, by applying the cold body externally instead of introducing it into the machine. It has been thought that a remarkable advantage might be secured by using the vapor of alcohol in that it possesses a stronger tension than the vapor of water at the same temperature. We can see in this only a fresh obstacle to be overcome. The principal defect of the vapor of water is its excessive tension at an elevated temperature; now this defect exists still more strongly in the vapor of alcohol. As to the relative advantage in a greater production of motive power—an advantage attributed to it—we know by the principles above demonstrated that it is imaginary.

It is thus upon the use of atmospheric air and vapor of water that subsequent attempts to perfect heat-engines should be based. It is to utilize by means of these agents the greatest possible falls of caloric that all efforts should be directed.

Finally, we will show how far we are from having realized, by any means at present known, all the motive power of combustibles.

One kilogram of carbon burnt in the calorimeter furnishes a quantity of heat capable of raising one degree Centigrade about 7000 kilograms of water, that is, it furnishes 7000 units of heat according to the definition of these units given on page 41.

The greatest fall of caloric attainable is measured by the difference between the temperature produced by combustion and that of the refrigerant bodies. It is difficult to perceive any other limits to the temperature of combustion than those in which the combination between oxygen and the combustible may take place. Let us assume, however, that 1000° may be this limit, and we shall certainly be below the truth. As to the temperature of the refrigerant, let us suppose it 0°. We estimated approximately (page 43) the quantity of motive power that 1000 units of heat develop between 100° and 99°. We found it to be 1.112 units of power, each equal to 1 cubic meter of water raised to a height of 1 meter.

If the motive power were proportional to the fall of caloric, if it were the same for each thermometric degree, nothing would be easier than to estimate it from 1000° to 0°. Its value would be

$$1.112 \times 1000 = 1112.$$

But as this law is only approximate, and as possibly it deviates much from the truth at high temperatures, we can only make a very rough estimate. We will suppose the number 1112 reduced one-half, that is, to 560.

Since a kilogram of carbon produces 7000 units of heat, and since the number 560 is relative to 1000 units, it must be multiplied by 7, which gives

$$7 \times 560 = 3920.$$

This is the motive power of 1 kilogram of carbon. In order to compare this theoretical result with that of experiment, let us ascertain how much motive power a kilogram of carbon actually develops in the best-known steam-engines.

The engines which, up to this time, have shown the best results are the large double-cylinder engines used in the drainage of the tin and copper mines of Cornwall. The best results that have been obtained with them are as follows:

65 millions of lbs. of water have been raised one English foot by the bushel of coal burned (the bushel weighing 88 lbs.). This is equivalent to raising, by a kilogram of coal, 195 cubic meters of water to a height of 1 meter, producing thereby 195 units of motive power per kilogram of coal burned.*

195 units are only the twentieth of 3920, the theoretical maximum; consequently $\frac{1}{20}$ only of the motive power of the combustible has been utilized.

We have, nevertheless, selected our example from among the best steam-engines known.

---

* The result given here was furnished by an engine whose large cylinder was 45 inches in diameter and 7 feet stroke. It is used in one of the mines of Cornwall called Wheal Abraham. This result should be considered as somewhat exceptional, for it was only temporary, continuing but a single month. Thirty millions of lbs. raised one English foot per bushel of coal of 88 lbs. is generally regarded as an excellent result for steam-engines. It is sometimes attained by engines of the Watt type, but very rarely surpassed. This latter product amounts, in French measures, to 104,000 kilograms raised one meter per kilogram of coal consumed.

According to what is generally understood by one horse power, in estimating the duty of steam-engines, an engine of ten horse-power should raise per second 10×75 kilograms, or 750 kilograms, to a height of one meter, or more, per hour; $750 \times 3600 = 2{,}700{,}000$ kilograms to one meter. If we suppose that each kilogram of coal raised to this height 104,000 kilograms, it will be necessary, in order to ascertain how much coal is burnt in one hour by our ten-horse power engine, to divide 2,700,000 by 104,000, which gives $\frac{2700}{104} = 26$ kilograms. Now it is seldom that a ten-horse power engine consumes less than 26 kilograms of coal per hour.

Most engines are greatly inferior to these. The old engine of Chaillot, for example, raised twenty cubic metres of water thirty-three metres, for thirty kilograms of coal consumed, which amounts to twenty-two units of motive power per kilogram—a result nine times less than that given above, and one hundred and eighty times less than the theoretical maximum.

We should not expect ever to utilize in practice all the motive power of combustibles. The attempts made to attain this result would be far more hurtful than useful if they caused other important considerations to be neglected. The economy of the combustible is only one of the conditions to be fulfilled in heat-engines. In many cases it is only secondary. It should often give precedence to safety, to strength, to the durability of the engine, to the small space which it must occupy, to small cost of installation, etc. To know how to appreciate in each case, at their true value, the considerations of convenience and economy which may present themselves; to know how to discern the more important of those which are only secondary; to balance them properly against each other, in order to attain the best results by the simplest means: such should be the leading characteristics of the man called to direct, to co-ordinate the labors of his fellow men, to make them co-operate towards a useful end, whatsoever it may be.

Appendix: Selections from the Posthumous Manuscripts of Carnot

After Sadi Carnot's death, a bundle of his papers was found, labelled "Notes on mathematics, physics and other subjects"; there are 23 loose sheets. They consist of rough notes in arbitrary order, lists of references and data, questions and speculations and experiments to be done, together with what almost seem to be finished drafts of complete paragraphs. It is not possible to assign a date to these notes, though it is certain that some of them were written at the same time as the *Reflections*. Raveau has suggested a chronological order based on the development of thought and language. This order has been followed here, but a selection of only the more interesting and significant notes has been made. In them one can follow the growth of Carnot's conviction that heat was equivalent to work, culminating in his estimation of the conversion factor and in the list of proposed experiments. These experiments are almost identical to those carried out much later by Joule—including the Joule–Kelvin throttling experiment. At the same time these papers reveal his growing doubts about the whole basis of his demonstrations in the *Reflections*.

Journal de Physique: experiments on the heat released by friction:

Plates of ice rubbed together by Davy.

Pneumatic pump with a little opening in the piston or at the circumference.

Rumford's experiment on the boring of metals, hammering of lead, tin, copper, boring of wood.

Entry of air into a vacuum.

The same into more or less rarified air.

Entry of air into the bell jar of a pneumatic machine.

These experiments would allow us to measure the changes of temperature undergone by gases through changes of volume, and further they would provide the means of comparing these changes with the quantities of motive power produced or consumed.

They could be carried out on gases other than ordinary atmospheric air.

Up to the present time the changes caused in the temperature of bodies by motion have been very little studied. This class of phenomena merits the attention of observers, however. When bodies are in motion, especially when that motion disappears or when it produces motive power, remarkable changes take place in the distribution of heat and perhaps in its quantity.

We will collect a few facts which exhibit this phenomenon most clearly.

The Collision of Bodies. We know that in the collision of bodies there is always expenditure of motive power. Only perfectly elastic bodies provide an exception, and none such are found in nature.

We also know that always in the collision of bodies there occurs a change of temperature, an elevation of temperature. We cannot, as did Berthollet, attribute the heat set free in this case to the reduction of the volume of the body; for when this reduction has reached its limit the liberation of heat should cease. This does not occur.

It is sufficient that the body changes form by percussion, without change of volume, to release heat.

If, for example, we take a cube of lead and strike it successively on each of its faces, there will always be heat liberated without sensible diminution in this release as long as the blows are continued with equal force. This does not occur when medals are struck. In this case the metal cannot change form after the first blows of the die, and the effect of the collision is not conveyed to the medal, but to the threads of the screw which are strained, and to its supports.

It would seem then that heat set free should be attributed to the friction of the molecules of the metal which change place relative to one another, that is that heat is set free just where the moving force is expended.

A similar remark will apply to the collision of two bodies of differing hardness—lead and iron, for instance. The first of these metals becomes very hot, while the second does not vary sensibly in temperature. But the motive power is almost wholly exhausted in changing the form of the first of these metals. We may also cite, as a fact of the same nature, the heat produced by the extension of a metallic rod just before it breaks. Experiment has proved that, other things being equal, the greater the elongation before rupture the more considerable the elevation of temperature.*

* [Unfortunately this essay is unfinished. E. M.]

Is it absolutely certain that steam having operated an engine and produced motive power can raise the temperature of the water of condensation as if it had been conducted directly into it?*

If the molecules of bodies are never in close contact with each other whatever be the forces which separate or attract them, there can never be either a gain or loss of motive power in nature. This power must be as unchangeable in quantity as matter. Then the direct re-establishment of equilibrium in caloric, and its re-establishment with production of motive power, would be essentially different from one another.

But how can one imagine the forces acting on the molecules, if these are never in contact with one another, if each one of them is completely isolated? To postulate a subtle fluid in between would only postpone the difficulty, because this fluid would necessarily be composed of molecules.

It appears very difficult to penetrate into the real nature of bodies. To avoid erroneous reasoning, it would be necessary to investigate carefully the source of our knowledge of the nature of bodies, their form, their forces; to see what the primitive notions are, to see from what impressions they are derived; to see how one is raised successively to the different degrees of abstraction.

Is heat the result of a vibratory motion of molecules? If this is so, quantity of heat is simply quantity of motive power. As long as motive power is used to produce vibratory movements, the quantity of heat must be unchangeable; which seems to follow from experiments in calorimeters; but when it passes into movements of sensible extent, the quantity of heat can no longer remain constant.

Can examples be found of the production of motive power without actual consumption of heat? It seems that we may find production of heat with consumption of motive power (re-entry of air into a vacuum, for example).

What is the cause of the production of heat in combinations of substances? What is radiant caloric?

Liquefaction of bodies, solidification of liquids, crystallization—are they not forms of combinations of integrant molecules?

---

* [Taken in conjunction with the description of the cycle of operations in the *Reflections*, this remark should dispose of the idea that Carnot *always* thought of caloric in the same way as we do entropy. E. M.]

Supposing heat due to a vibratory movement, how can the passage from the solid or the liquid to the gaseous state be explained?*

When motive power is produced by the passage of heat from the body A to the body B, is the quantity of this heat which arrives at B (if it is not the same as that which has been taken from A, if a portion has really been consumed to produce motive power) the same whatever may be the substance employed to realize the motive power?

Is there any way of using more heat in the production of motive power, and of causing less to reach the body B? Could we even utilize it entirely, allowing none to go to the body B? If this were possible, motive power could be created without consumption of fuel, and by mere destruction of the heat of bodies.

What becomes of the heat given out by the earth and other astral bodies by means of radiation?

If, as mechanics seems to prove, there cannot be any real creation of motive power, then there cannot be any destruction of this power either—for otherwise all the motive power of the universe would end by being destroyed—hence there cannot be any real collision between bodies.

When air enters into a vacuum, it is its passage through a little opening and the movement imparted to it in the interior which seem to produce the rise of temperature.

We may be allowed to express here a hypothesis concerning the nature of heat.

At present, light is generally regarded as the result of a vibratory movement of the etherial fluid. Light produces heat, or at least accompanies the radiant heat and moves with the same velocity as heat. Radiant heat is therefore a vibratory movement. It would be ridiculous to suppose that it is an emission of matter while the light which accompanies it could only be a movement.

Could a motion (that of radiant heat) produce matter (caloric)? Undoubtedly no; it can only produce a motion. Heat is then the result of a motion.

Then it is plain that it could be produced by the consumption of motive power and that it could produce this power.

---

* [It should be remembered that gases were considered to be highly rarified solids in which the attractive forces between molecules had somehow changed into repulsive ones; the pressure exerted by a gas was due to the static effect of these repulsions. E. M.]

All the other phenomena—composition and decomposition of bodies, passage to the gaseous state, specific heat, equilibrium of heat; its more or less easy transmission, its constancy in calorimeter experiments—could be explained by this hypothesis. But it would be difficult to explain why, in the development of motive power by heat, a cold body is necessary; why motion cannot be produced by consuming the heat of a warm body.

Experiments to be made on heat and motive power:*

To repeat Rumford's experiments on the drilling of a metal in water, but to measure the motive power consumed at the same time as the heat produced; same experiments on several metals and on wood.

To agitate water vigorously in a small cask or in a double-acting pump having a piston pierced with a small opening.

Experiment of the same sort on the agitation of mercury, alcohol, air and other gases. To measure the motive power consumed and the heat produced.

To strike a piece of lead in several directions, to measure the motive power consumed and the heat produced. Same experiments on other metals.

To admit air into a vacuum or into air more or less rarified; the same for other gases or vapors. To examine the elevation of temperature by means of a manometer and a Breguet thermometer. Estimation of the error of the thermometer from the time required for the air to vary a certain number of degrees.†

Gay-Lussac's experiment with two equal vessels, one empty and the other filled with air, put into communication with one another.

Expel the air from a large reservoir in which it is compressed and check its velocity in a large pipe in which solid bodies have been placed; measure the temperature when it has become uniform. See if it is the same as in the reservoir. Same experiment with other gases and with vapor formed under different pressures.

---

* [There are several of these lists in the manuscript. These have been amalgamated here. E. M.]

† [In a later note, Carnot worked out the details of this experiment at considerable length, designing a quickly responding air thermometer to measure the instantaneous temperature of the gas. He was dissatisfied with Gay-Lussac's experiment (the precursor of Joule's experiment with the two cylinders) and the discrepancy between the small rises and falls of temperature in the two vessels compared with those expected from adiabatic compressions and expansions. E. M.]

To repeat Dalton's experiments and carry them on to pressures of thirty or forty atmospheres. To measure the constituent heat of the steam within these limits.

The same on the vapor of alcohol, of ether, of essence of turpentine, of mercury, to test whether the agent employed makes any difference in the production of motive power.

The same on water charged with a deliquescent salt—calcium chloride, for instance.

Is the law of tensions always the same? To measure the specific heat of steam.

Experiments to be done on the tension of vapors:

A graduated capillary tube filled with water, mercury or with oil and air. Plunge this tube into a bath of oil, of mercury or of molten lead. To measure the temperature with an air thermometer.

Same experiments with alcohol, ether, sulfide of carbon, muriatic ether, essence of turpentine, sulfur, phosphorus.

Experiments on the tension of steam with a boiler, and a thermometer tube full of air. A thermometer placed in a tube immersed in the boiler, open outwards and filled with oil or mercury.

Experiments using a simple capillary tube filled in three parts one after the other—first of air, second of mercury, third of water or other liquid whose tension can be measured (of alcohol, ether, essence of turpentine, lavender, sulfide of carbon, muriatic ether, etc.). One end of the tube may be immersed in a bath of mercury or oil, the temperature of which is to be measured. The column of mercury can be made long enough to allow the air to be previously compressed or rarified.

The tube bent into a spiral at one end, the straight part being graduated (thus permitting the tension of mercury vapor to be measured).

Experiments on the tension of vapors at low temperature, with a thermometer tube bent round and filled partly with mercury, partly with water or alcohol. The mercury will act because of its weight, the upper part of the tube will be empty and sealed or fully open to the atmosphere. The bulb immersed in water whose temperature

is to be measured. If the tube is sealed the upper part must be cooled.

The bulb might contain water, ether or essence of turpentine.

If the tube is sealed, the tension of mercury vapor could be measured.

Experiments on the constituent heat of vapors by means of a barometer tube having two enlarged bulbs. One of the bulbs may

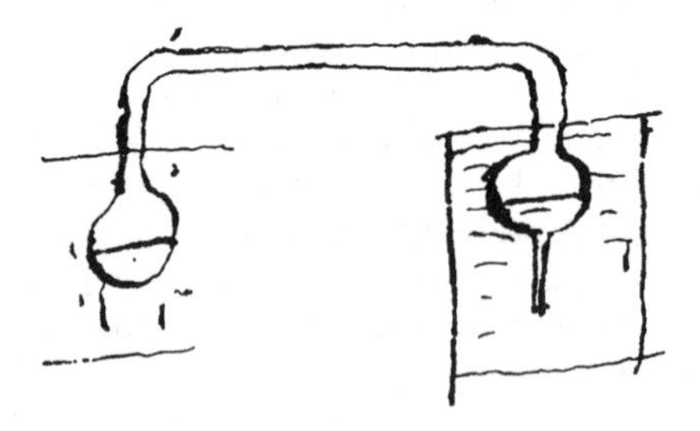

be immersed in cold water and the elevation of temperature of this water will indicate the constituent heat of the vapor.

The other bulb may be warmed either by boiling liquid or by fire.

Water, alcohol, turpentine, ether, mercury, acetic acid, sulfide of carbon.

The operation may be repeated and the results added.

Instruments: good thermometer, 1; ordinary blown tubes, 6; the same, longer, 3; blowpipes* with second ball blown, 6; ordinary blowpipes, 2. Alcohol, ether, turpentine, sulfide of carbon, 10

* [A blowpipe consisted essentially of a tube with a bulb at the end. E. M.]

grams, linseed oil, 1 kilogram, tin-plate vessel for heating it. Small portable furnaces.

If it were found that gases do not change in temperature when they expand without producing motive power—when, for example, they escape from a constant pressure through a small opening into another

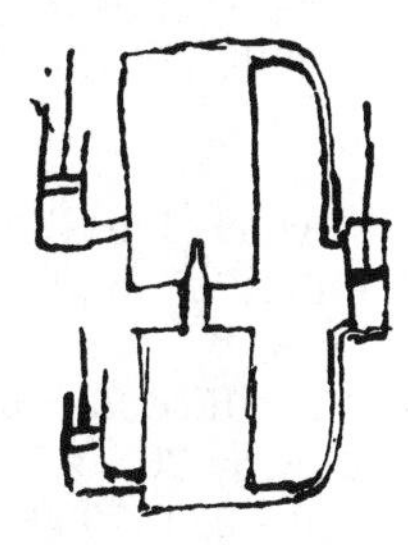

vessel where they are subjected to a smaller pressure which is also constant—it would follow

(1) that the consumption of motive power produces caloric;
(2) that the quantity produced is exactly that given up by the gas after its reduction of volume.

Indeed, if a gas is compressed to half, for example, keeping the temperature constant, and is then allowed to escape through a small opening into a vessel where the pressure is maintained at its original value, this gas would be brought back at the end of the operation, precisely to its original state; a quantity of motive power will have been consumed which is equal to that necessary to reduce the gas to half its volume.

Heat is simply motive power, or rather motion which has changed its form. It is a movement among the particles of bodies. Where-ever there is destruction of motive power, there is at the same time production of heat in quantity exactly proportional to the quantity of motive power destroyed. Reciprocally, wherever there is destruction of heat, there is production of motive power.

We can establish the general proposition that motive power is, in quantity, invariable in nature; that it is, correctly speaking, never either produced or destroyed. It is true that it changes its form—that is, it produces sometimes one sort of motion, sometimes another—but it is never annihilated.

This principle follows straightforwardly, as it were, from mechanics; indeed, reasoning shows that there cannot be loss of *vis viva* or, what is the same thing, of motive power if bodies act upon each other without directly touching one another, without actual collision. Now everything leads us to believe that the molecules of bodies are always separated from one another by some space, that they are never actually in contact. If they touched one another they would remain united and consequently change form.

According to some ideas which I have formed on the theory of heat, the production of a unit of motive power necessitates the destruction of 2.70 units of heat.*

A machine which would produce 20 units of motive power per kilogram of coal ought to destroy $20 \times 2.70/7000$ of the heat developed by the combustion. $20 \times 2.70/7000 = 8/1000$ roughly, that is less than $1/100$.

When a hypothesis no longer suffices to explain phenomena, it should be abandoned.

This is the case with the hypothesis which regards caloric as matter, as a subtle fluid.

The experimental facts tending to destroy this theory are as follows:

(1) The development of heat by percussion or the friction of bodies (Rumford's experiments, friction of wheels on their spindles, on the axles, experiments to be made). Here the elevation of temperature takes place at the same time in the body rubbing and in the body rubbed. Moreover, they do not change perceptibly in form or nature (to be proved). Thus heat is produced by motion.

* [The unit of work is the kilogram-meter, so that this corresponds in modern language to $J = 3.7$ joules/calorie. Unfortunately Carnot has left no account of his method of deduction. Two attempts have been made to interpret his figure. L. Decombe (*Comptes Rendus* CLXVIII, 268 (1919)) suggested he may have interpreted the figures for $C_p - C_v$ for air as equivalent to the work of expansion—as Mayer did much later. On the other hand C. Raveau (*Comptes Rendus* CLXVIII, 549 (1919)) suggested a more roundabout method. In the *Reflections* Carnot calculated the area of a cycle bounded by 0° and 0°.001 isotherms as 3.72×10^{-7} units of work, and the heat of compression as 0.267 units of heat. He may then have written the area of the cycle $dp \cdot dV$ as equal to $(dp/dT)dT \cdot dV$ which for a perfect gas is equal to $(dT/T)p \cdot dV$. $p \cdot dV$ is the work of compression, so that putting $dT = 0°.001$ and $T = 267°$ the value of the mechanical equivalent follows. E. M.]

If it is matter, it must be admitted that the matter is created by motion.

(2) When an air pump is worked and at the same time air is admitted into the receiver, the temperature remains constant in the receiver. It remains constant on the outside. Consequently the air compressed by the pump must rise in temperature above the air outside, and it is expelled at a higher temperature. The air enters at a temperature of 10°, for instance, and leaves at another, 10°+90° or 100°, for example. Thus heat has been created by motion.

(3) If the air in a reservoir is compressed, and at the same time allowed to escape through a little opening, there is a rise of temperature from the compression and a lowering of temperature from the escape (according to Gay-Lussac and Welter). The air then enters at one side at one temperature and escapes at the other side at a higher temperature, from which follows the same conclusion as in the preceding case.

(Experiment to be done: To fit to a high-pressure boiler a cock and a tube leading to it and emptying into the atmosphere; to open the cock a little way, and present a thermometer to the outlet of the steam; to see if it remains at 100° or more; to see if the steam is liquefied in the pipe; to see whether it comes out cloudy or transparent.)

(4) The elevation of temperature which takes place at the time of the entry of air into vacuum, an elevation that cannot be attributed to the compression of the air remaining (air which may be replaced by steam), can therefore be attributed only to the friction of the air against the walls of the opening, or against the interior of the receiver, or against itself.

(5) Gay-Lussac showed (it is said) that if two vessels were put in communication with each other, the one a vacuum, the other full of air, the temperature would rise in one as much as it would fall in the other. If afterwards both be compressed one half, the second would return to its original temperature and the first to a much higher temperature. Mixing them, the whole mass would be heated.

MEMOIR ON THE MOTIVE POWER OF HEAT

BY É. CLAPEYRON

Mining Engineer

(*Journal de l'École Polytechnique*, XIV, [1834] 153
and Poggendorff's *Annalen der Physik*, LIX, [1843] 446, 566)

Translated and edited by
E. MENDOZA

Émile Clapeyron. (*From the Archives of the Académie des Sciences.*)

Memoir on the Motive Power of Heat

§ I

THERE are few questions more worthy of the attention of theoreticians and physicists than those which refer to the constitution of gases and vapors; the part which they play in nature, and the uses that industry puts them to, explain the many important investigations on them; but this vast question is far from being exhausted. Mariotte's law and Gay-Lussac's law state the relations which exist between the volume, pressure and temperature of a given quantity of gas; both have been accepted by scientists for a long time. The recent experiments carried out by Arago and Dulong leave no doubt as to the exactness of the first between very wide limits of pressure; but these results give no indication of the quantity of heat possessed by the gases or set free from them by pressure or lowering of temperature, neither do they give the law of specific heats at constant pressure and constant volume. This part of the theory of heat has nevertheless been the subject of careful investigations, among which may be mentioned the work of Delaroche and Bérard on the specific heats of gases. Lastly, in a memoir which he published under the title *Researches on the Specific Heats of Elastic Fluids*, Dulong established, by experiments which are free from all objections, that *equal volumes of all elastic fluids at a given temperature and pressure, compressed or expanded suddenly by a given fraction of their volumes, release or absorb the same absolute quantity of heat.**

Laplace and Poisson have published remarkable theoretical studies on this subject which however are based on hypotheses which can be disputed; they agree that the ratio of the specific heat at constant volume to that at constant pressure does not vary, and that the quantities of heat absorbed by gases are proportional to their temperatures.

Among studies which have appeared on the theory of heat I will mention finally a work by S. Carnot, published in 1824, with the

* [On the caloric theory, the heat set free when a given gas was compressed depended only on the initial and final volumes; it was the same whether the compression was isothermal or adiabatic. E. M.]

title *Reflections on the Motive Power of Fire.* The idea which serves as a basis of his researches seems to me to be both fertile and beyond question; his demonstrations are founded on *the absurdity of the possibility of creating motive power or heat out of nothing.* Here are statements of several theorems to which this new method of reasoning leads.

1. *When a gas passes, without changing its temperature, from a given volume and pressure to another given volume and pressure, the quantity of caloric absorbed or set free is always the same whatever the nature of the gas chosen for the experiment.*

2. *The difference between the specific heats under constant pressure and constant volume is the same for all gases.*

3. *When a gas varies its volume, without changing its temperature, the quantities of heat absorbed or set free by the gas are in arithmetic progression if the increases or decreases of volume are in geometric progression.*

This new method of demonstration seems to me worthy of the attention of theoreticians; it seems to me to be free of all objection, and it has acquired a new importance since the verification found in the work of Dulong, who has demonstrated experimentally the first theorem whose enunciation I have just recalled.

I believe that it is of some interest to take up this theory again; S. Carnot, avoiding the use of mathematical analysis, arrives by a chain of difficult and elusive arguments at results which can be deduced easily from a more general law which I shall attempt to prove. But before starting on the subject it is useful to restate the fundamental axiom which serves as a basis for Carnot's researches, and which will also be my starting point.

§II

It has been known for a long time that heat can develop motive power and, conversely, that by means of motive power heat can be produced. In the first case, it must be observed that a certain quantity of caloric always passes from one body at a given temperature to another body at a lower temperature; thus in steam engines, the production of mechanical force is accompanied by the passage of a part of the heat of combustion developed in the furnace whose temperature is very high, to the water of the condenser, whose temperature is very much lower.

Conversely it is always possible to use the passage of caloric from a hot body to a cold one to produce a mechanical force; to do this it is only necessary to construct a mechanism like an ordinary steam-engine, where the hot body generates the steam and the cold body acts as condenser.

The result is that there is a loss of *vis viva*, of mechanical force or quantity of action, whenever there is an immediate contact between two bodies at different temperatures and heat passes straight from one to another; therefore, in any mechanism designed to produce motive power from heat, there is a loss of force whenever there is a direct communication of heat between two bodies at different temperatures, and it follows that the maximum effect can be produced only by a mechanism in which contact is made only between bodies at equal temperatures.

Now our knowledge of the theory of gases and vapors shows how this can be achieved.

Let us imagine two bodies, one maintained at a temperature T, the other at a lower temperature t, such as for example the walls of a boiler in which the heat developed by combustion continually replaces that which the steam takes away; and the condenser of an ordinary heat-engine, in which a current of cold water all the time removes the heat given up by the steam by condensation and that due to its own temperature. For simplicity we will call the first body A, the second B.

Then, let us take any gas whatever at temperature T and let us put it in contact with the source A of heat; let us represent its volume v_0 by the abscissa AB and its pressure by the ordinate CB (Fig. 1).

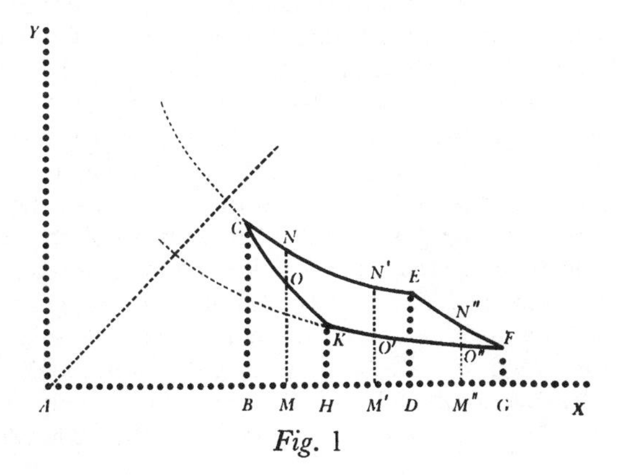

Fig. 1

If the gas is enclosed in a deformable vessel and is allowed to expand in an empty space where it can lose heat neither by radiation nor by contact, the source *A* of heat will at all times provide the quantity of caloric which its increase of volume causes to become latent, and it will keep the same temperature *T*. Its pressure however will diminish following Mariotte's law. The law of this variation can be represented geometrically by the curve *CE* where the abscissae are the volumes, and the ordinates the corresponding pressures.

Let us suppose that the expansion of the gas is continued till the original volume *AB* has become *AD*; and let *DE* be the pressure corresponding to the new volume; the gas will have developed a quantity of mechanical action during its expansion given by the integral of the product of the pressure times the differential of the volume, represented geometrically by the area contained between the axis of abscissae, the two co-ordinates *CB*, *DE*, and the portion *CE* of the hyperbola.

Let us suppose now that the body *A* is removed and that the expansion of the gas continues inside an envelope impermeable to heat; then since a part of its perceptible caloric becomes latent its temperature drops and its pressure decreases more rapidly according to an unknown law, which can be represented geometrically by a curve *EF* whose abscissae are the volume of the gas and whose ordinates are the corresponding pressures; we will suppose that the expansion of the gas is continued till the successive reductions of the perceptible caloric of the gas have brought it from the temperature *T* of the body *A* to the temperature *t* of the body *B*. Its volume is therefore *AG*, and the corresponding pressure *FG*.

It will be seen that the gas, during this second part of its expansion, develops a quantity of mechanical action represented by the area of the mixtilinear trapezium *DEFG*.

Now that the gas has been brought to the temperature *t* of the body *B*, let us bring the two into contact; if the gas is compressed in an envelope impermeable to heat, but in contact with the body *B*, the temperature of the gas tends to rise because of the release of latent caloric made perceptible by the compression, but as it is produced it is absorbed by the body *B* so that the temperature of the gas remains equal to *t*. As a result, the pressure increases according to Mariotte's law; it will be represented geometrically by the ordinates of a hyperbola *KF* and the corresponding abscissae will represent the volumes. Let us suppose that the compression is continued till the heat released by the compression of the gas and absorbed by the body *B* is exactly equal to the heat communicated by the source *A* to the gas, during its expansion in contact with it in the first part of the operation.*

Then let *AH* be the volume of the gas and *HK* the corresponding pressure. In this state the gas possesses the same absolute quantity of heat as at the start of the operation, when it occupied the volume *AB* under pressure *CB*. If then the body *B* is removed and the gas is further compressed inside an envelope which is impermeable to heat until the volume *AH* becomes the volume *AB*, its temperature increases all the time by the release of latent caloric made perceptible by the compression. At the same time the pressure increases, and when the volume is reduced to *AB* the temperature returns to *T* and the pressure to *BC*. Now the different states in which a given mass of gas can exist are characterized by the volume, pressure, temperature, and the absolute quantity of heat which it contains; if two of these four quantities are known, the other two are determined; thus, in the case under discussion, since the absolute quantity of heat and the volume are the same as they were at the start of the operation, it is certain that the temperature and the pressure will also be what they were then. Consequently the unknown law, of how the pressure varies when the volume of the gas is reduced inside its impermeable envelope, is represented by a curve *KC* which passes through point *C*, and whose abscissae and ordinates always represent volumes and pressures.

However, the reduction of the volume of the gas from *AC* to *AB* will have consumed a quantity of mechanical action which, by the same arguments which we have given above, will be represented by the two mixtilinear trapeziums *FGHK* and *KHBC*. If we subtract these two trapeziums from the first two *CBDE* and *EDFG* which represent the quantity of action developed during the expansion of the gas, the difference, which will be equal to the kind of curvilinear parallelogram *CEFK*, will represent the quantity of action developed by the cycle of operations just described, at the end of which the gas is in precisely its original state.

However, all the heat given up by the body *A* to the gas during its expansion in contact with it has flowed into the body *B* during the compression of the gas which took place in contact with that body.

* [Here "heat" should be taken to mean entropy. E. M.]

Here, therefore, mechanical force has been developed by the passage of caloric from a hot body to a cold body, and this passage has been accomplished without any contact between bodies at different temperatures.

The reverse operation is equally possible; thus let us take the same volume of gas AB at the temperature T and at pressure BC; let us enclose it in an envelope impermeable to heat, and dilate it so that its temperature falls gradually till it is equal to t; let us continue the expansion in the same envelope, but after having introduced the body B which has the same temperature; this provides the gas with the heat necessary to maintain its temperature and we continue the operation till the body B has given the gas the heat which it had received from it in the previous operation. Next let us remove the body B and compress the gas in an impermeable envelope till its temperature becomes once again equal to T. Then let us bring up the body A which has the same temperature, and continue to reduce the volume till all the heat taken from the body B has been given up to the body A. The gas then must have the same temperature and must possess the same absolute quantity of heat as at the beginning of the operation; whence it must occupy the same volume and must be at the same pressure.

Here the gas passes successively, but in the reverse order, through all the states of temperature and pressure through which it passed in the first series of operations; as a result the expansions have become compressions and vice versa; but they follow the same law. It follows that the quantities of action developed in the first case are absorbed in the second and vice versa, but they keep the same numerical values, because the elements of the integrals which compose them are the same.

Thus it can be seen that in causing heat to pass from a body maintained at a given temperature, using the method first described, a certain quantity of mechanical action is developed which is equal to that which must be used up to cause the same quantity of heat to pass from the cold body to the hot body by the reverse procedure just described.

A similar result can be arrived at through the vaporization of any liquid. Let us take such a liquid and put it in contact with the body A in a rigid envelope impermeable to heat; we suppose the temperature of the liquid to be equal to the temperature T of the body A. We mark on the axis of the abscissae AX (Fig. 2) a quantity AB equal to the volume of liquid and, on a line parallel to the axis of ordinates AY, a quantity BC equal to the vapor pressure of the liquid corresponding to the temperature T.

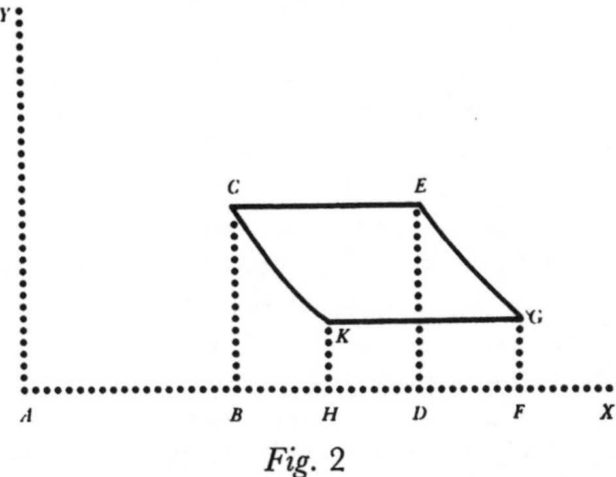

Fig. 2

If we increase the volume of the liquid, a part of it passes into the vapor state, and as the source of heat A provides the latent caloric necessary for its formation, the temperature remains constant and equal to T. If we mark on the axis of the abscissae quantities representing the successive volumes which the mixture of liquid and vapor occupy, and if the corresponding values of the pressure are taken for ordinates, then as the pressure remains constant the pressure curve is reduced here to a straight line parallel to the axis of the abscissae.

When a certain quantity of vapor has been formed and when the mixture of liquid and vapor occupies a volume AD, the body A is removed and the expansion continued. Then a further quantity of liquid passes to a gaseous state and a part of the perceptible caloric becomes latent, the temperature of the mixture falls as well as the pressure; let us suppose that the dilation is continued till the temperature, after falling gradually, becomes equal to the temperature t of the body B; let AF be the volume, FG the corresponding pressure. The law of the variation of pressure is given by some curve EG which passes through the points E and G.

During this first part of the operation a quantity of action will have been developed represented by the areas of the rectangle $BCED$ and the mixtilinear trapezium $EGFD$.

Let us now bring up the body B, put it in contact with the mixture of liquid and vapor and gradually reduce the volume; a part of the vapor passes to the liquid state and since the latent heat which it releases while condensing is absorbed by the body B as it is produced, the temperature remains constant equal to t. We continue to reduce the volume in this way till all the heat provided by the body A in the first part of the operation has been given to the body B.

Let AH be the volume then occupied by the mixture of vapor and liquid; the corresponding pressure is KH equal to GF; since the temperature remains equal to t during the reduction of the volume from AF to AH, the law of pressure between these two limits is represented by the line KG parallel to the axis of the abscissae.

Having arrived at this point, the mixture of vapor and liquid with which we are working, which occupies the volume AH under the pressure KH, and at a temperature t, possesses the same absolute quantity of heat that the liquid possessed at the start of the operation; if, therefore, the body B is removed and the condensation is continued in a vessel impermeable to heat, till the volume becomes again equal to AB, the same quantity of matter is occupying the same volume and possesses the same quantity of heat as at the beginning of the operation; its temperature and pressure must therefore be the same as at that stage; the temperature must also become again equal to CB; the law of pressures during this last part of the operation will therefore be given by a curve passing through the points K and C and the quantity of action absorbed during the reduction of the volume AF to AB is represented by the rectangle $FHKG$ and the mixtilinear trapezium $BCKH$.

If, therefore, the quantity of action developed during the expansion is subtracted from that which is absorbed during the compression, there remains the difference, the area of the mixtilinear parallelogram $CEGK$, which represents the quantity of action developed during the complete series of operations just described, at the end of which the liquid finds itself in its original state.

But it must be noted that all the caloric communicated by the body A has passed into the body B and that this transfer has been carried out without there having been any contact other than that between bodies at the same temperatures.

In the same way as for gases, it may be proved that by repeating the same operation in the reverse order, heat may be made to pass from the body B to the body A, but that this result can only be achieved by the absorption of a quantity of action equal to that developed by the passage of the same quantity of caloric from the body A to the body B.

From what has gone before, it follows that a quantity of mechanical action, and a quantity of heat which can pass from a hot body to a cold body, are quantities of the same nature, and that it is possible to replace the one by the other; in the same manner as in mechanics a body which is able to fall from a certain height and a mass moving with a certain velocity are quantities of the same order, which can be transformed one into the other by physical means.*

Hence it also follows that the quantity of action F developed by the passage of a certain quantity of heat C from a body maintained at a temperature t, by one of the procedures just outlined, is the same whatever the gas or liquid employed, and is the greatest which it is possible to attain. For suppose that by any process whatever the quantity of heat C were made to pass from the body A to the body B and that it was possible to produce a greater quantity of action F', we could use a part F to restore the quantity of heat C from the body B to the body A by one of the two methods just described; the *vis viva* F used for this purpose would, as we have seen, be equal to that developed by the passage of the same quantity of heat C from the body A to the body B; so by hypothesis it is smaller than F', so there would be produced a quantity of action $F'-F$, created out of nothing and without consumption of heat, an absurd result which would lead to the possibility of creating force or heat gratuitously and without limit. It seems to me that the impossibility of such a result can be accepted as a fundamental axiom of mechanics; no one has ever objected to Lagrange's demonstration of the principle of virtual velocities using pulleys, and this seems to me to depend on something similar.

In the same way it can be shown that no gas or vapor exists which, if used to transmit heat from a hot body to a cold one by the methods described, can develop a quantity of action greater than any other gas or vapor.

We therefore base our researches on the following principles: Caloric, passing from one body to another maintained at a smaller temperature, can give rise to the production of a certain quantity of mechanical action; there is a loss of *vis viva* whenever there is contact between bodies at different temperatures. The maximum effect is produced when the passage of the caloric from the hot to the cold body is effected by one of the methods just described. In

* [This extraordinary paragraph is an unambiguous statement of the First Law of Thermodynamics. It serves to emphasize the point made in the Introduction, that the caloric theory and the *vis viva* theory were not regarded as mutually exclusive. E. M.]

addition this is independent of the chemical nature of the liquid or gas employed, of its quantity or its pressure; so that the maximum quantity of action which the passage of a given quantity of heat from a cold to a hot body can develop is independent of the nature of the agents used.

§ III

We will now translate analytically the various operations described in the previous paragraph; we will deduce the expression for the maximum quantity of action developed by the passage given quantity of heat from a body maintained at a certain temperature to a body maintained at a lower temperature, and we will arrive at some new relations between the volume, pressure, temperature and absolute heat or latent caloric of solids, liquids or gases.

Let us again consider the two bodies *A* and *B* and suppose that the temperature of the body *B* is smaller than the temperature *t* of the body *A* by an infinitely small amount *dt*. Let us first suppose that it is a gas which serves to transmit the caloric from the body *A* to the body *B*. Let V_0 be the volume of the gas at the pressure p_0 and at the temperature t_0; let *p* and *v* be the volume and the pressure of the same mass of the gas at the temperature *t* of the body *A*. Mariotte's law, combined with that of Gay-Lussac, establishes between these different quantities the relation

$$pv = \frac{p_0 v_0}{267+t_0}(267+t),$$

or simply

$$\frac{p_0 v_0}{267+t_0} = R:$$

$$pv = R(267+t).$$

The body *A* is put into contact with the gas. Let $me = v$, $ae = p$ (Fig. 3). If the gas is dilated by an infinitesimal amount $dv = eg$, the temperature remains constant because of the presence of the source of heat *A*; the pressure diminishes and becomes equal to the ordinate *bg*. Now the body *A* is removed and the gas dilated in an envelope impermeable to heat, by an infinitesimal amount *gh*, till the heat which has become latent lowers the temperature of the gas by an infinitesimal amount dt and thus brings it to the temperature $t-dt$ of the body B. As a result of this lowering of temperature

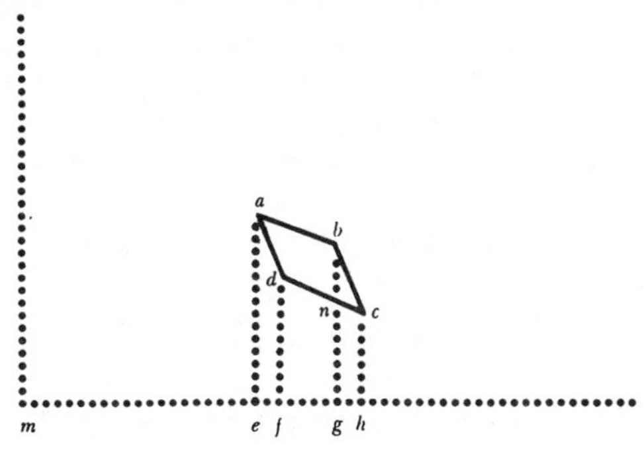

Fig. 3

the pressure falls more rapidly than in the first part of the operation and becomes *ch*. We now bring up the body B and reduce the volume *mh* by an infinitesimal amount *fh*, so calculated that during this compression the gas gives up to the body B all the heat which it drew from the body A during the first part of the operation. Let *fd* be the corresponding pressure; having done that we remove the body B and continue to compress the gas till it has regained the volume *me*. Then the pressure has become again equal to *ae* as demonstrated in the previous paragraph, and it can also be shown in the same way that the quadrilateral *abcd* measures the quantity of action produced by the transfer to the body B of the heat drawn from the body A during the dilation of the gas.

Now it is easy to show that this quadrilateral is a parallelogram; this is a result of the infinitely small values given to the variations of volume and pressure: for imagine that through each of the points in the plane on which the quadrilateral *abcd* is drawn, we erect perpendiculars to this plane, and that on each of them are marked two quantities T and Q as distances from the plane, the first equal to the temperature, the second to the absolute quantity of heat which the gas possesses when the volume and pressure have the values assigned to them by the abscissa v and ordinate p corresponding to each point.

The lines *ab* and *cd* belong to the projections of two equal-temperature curves on the temperature surface passing through two points infinitely close together; *ab* and *cd* are therefore parallel; *ad* and *bc* are also the projections of two curves for which $Q=\text{const.}$ on the surface $Q=f(p, v)$ which also passes through two infinitely close points, hence these two elements are also parallel; the quadrilateral *abcd* is therefore a parallelogram and it is easy to see that its area is found by multiplying the variation of volume during the contact of the gas with the body A or the body B, that is *eg*, or *fh* which is equal to it, by the difference *bn* of the pressure during these two operations and corresponding to the same value of the volume v. Now since *eg* and *fh* are the differentials of the volume, they are equal to dv; *bn* is obtained by differentiating the equation $pv=R(267+t)$ holding v constant; therefore

$$bn = dp = R\frac{dt}{v}.$$

The quantity of action developed is therefore expressed as

$$R\frac{dt \cdot dv}{v}.$$

We still have to determine the quantity of heat needed to produce this effect; it is equal to that which the gas drew from the body A while its volume increased by dv keeping the same temperature; now since Q is the absolute quantity of heat which the gas possesses, it must be a certain function of p and v, taken as independent variables; the quantity of heat absorbed by the gas is therefore

$$dQ = \frac{dQ}{dv} \cdot dv + \frac{dQ}{dp} \cdot dp,$$

but the temperature remains constant during the variation of volume so

$$v \cdot dp + p \cdot dv = 0, \qquad \text{whence} \qquad dp = -\frac{p}{v} \cdot dv,$$

and hence

$$dQ = \left(\frac{dQ}{dv} - \frac{p}{v} \cdot \frac{dQ}{dp}\right) dv.$$

Dividing the effect produced by this value of dQ, we have

$$\frac{Rdt}{v\frac{dQ}{dv} - p\frac{dQ}{dp}}$$

as the expression for the maximum effect which a quantity of heat equal to unity can develop by passing from a body maintained at the temperature t to a body maintained at the temperature $t-dt$.

We have demonstrated that this quantity of action developed is independent of the agent which serves to transmit the heat; it is therefore the same for all gases, neither does it depend on the mass of the body employed; but there is nothing to show that it is independent of the temperature;

$$\left(v\frac{dQ}{dv}-p\frac{dQ}{dp}\right)$$

must therefore be equal to an unknown function of t which is the same for all gases.

Now because of the equation $pv=R(267+t)$, t is itself a function of the product pv, therefore we have the partial differential equation

$$v\frac{dQ}{dv}-p\frac{dQ}{dp} = F(p\cdot v);$$

this has for integral

$$Q = f(p\cdot v)-F(p\cdot v)\log_e p.$$

The generality of this formula is not at all changed by replacing these two arbitrary functions of the product pv by the functions B and C of the temperature, multiplied by the coefficient R; therefore we have

$$Q = R(B-C\log p).^*$$

It is easy to verify that this value of Q satisfies all the required conditions; for we have

$$\frac{dQ}{dv} = R\left(\frac{dB}{dt}\cdot\frac{p}{R}-\log p\cdot\frac{dC}{dt}\cdot\frac{p}{R}\right)$$

$$\frac{dQ}{dp} = R\left(\frac{dB}{dt}\cdot\frac{v}{R}-\log p\cdot\frac{dC}{dt}\cdot\frac{v}{R}-C\frac{1}{p}\right);$$

from this is found

$$v\frac{dQ}{dv}-p\frac{dQ}{dp} = CR$$

and consequently

$$\frac{Rdt}{v\frac{dQ}{dv}-p\frac{dQ}{dp}}=\frac{dt}{C}.$$

* [This corresponds to an integrated form of the equation for a reversible change with a perfect gas:

$$T\cdot dS = C_p\cdot dT-\frac{RT}{J}\cdot\frac{dp}{p}$$

where R is expressed in mechanical units and J is the mechanical equivalent of heat. The function C is in fact equal to T/J. E. M.]

The function C which multiplies the logarithm of the pressure in the value of Q is of great importance, as can be seen; it is independent of the nature of the gas and is a function of the temperature only; it is essentially positive and serves as a measure of the maximum quantity of action which heat can develop.

We have seen that of the four quantities Q, t, p and v, if two are known the other two are determined; they must therefore be connected by two equations; one of them,

$$pv = R(267+t),$$

results from the laws of Mariotte and Gay-Lussac combined. The equation

$$Q = R(B-C\log p),$$

which we deduce from our theory, is the second. However, the numerical calculation of the changes which gases undergo when the volume and pressure are varied arbitrarily, demands a knowledge of the functions B and C.

We will see later that approximate values of the function C can be obtained over a considerable range of temperatures; further, having been obtained for one gas it will be the same for all. As to the function B, it can vary from one gas to another; however, it is probable that it is the same for all simple gases: at least, that seems to follow from experiment, from the fact that they have the same heat capacities.

Let us return to the equation

$$Q = R(B-C\log p).$$

Let us compress a gas occupying volume v at pressure p till the volume becomes v', and let it be cooled till the temperature returns to the same point. Let p' be the new value of the pressure; let Q' be the new value of Q; we have

$$Q-Q' = RC\log\frac{p'}{p} = RC\log\frac{v}{v'}.$$

Since the function C is the same for all gases, it is seen that *equal volumes of all elastic fluids, taken at the same temperature and under the same pressure, when compressed or expanded by the same fraction of their volumes, set free or absorb the same absolute quantity of heat.* This is the law deduced by Dulong by direct experiment.

In addition this equation shows that *when a gas varies its volume without changing its temperature, the quantities of heat absorbed or set free are in arithmetic progression if the increments or reductions of volume are in geometrical progression.* This result has also been enunciated by Carnot in the work cited.

The equation

$$Q-Q' = RC\log\left(\frac{v}{v'}\right)$$

expresses a more general law; it takes account of all the circumstances which can influence the phenomenon, such as the pressure, the volume and the temperature.

For since

$$R = \frac{p_0v_0}{267+t_0} = \frac{pv}{267+t},$$

we have

$$Q-Q' = \frac{pv}{267+t}C\log\frac{v}{v'}.$$

This equation shows the influence of pressure; it shows that *equal volumes of all gases, taken at the same temperature, if compressed or expanded by a given fraction of their volumes, set free or absorb quantities of heat proportional to the pressure.*

This result explains how the sudden entry of air into the receiver of the pneumatic pump does not release a perceptible quantity of heat. The vacuum of the pneumatic pump is nothing more than a volume v of gas whose pressure p is very small; if atmospheric air is allowed to enter, its pressure p becomes suddenly equal to that of the atmosphere, p', its volume is reduced to v' and the expression for the heat released is

$$C\frac{pv}{267+t}\log\frac{v}{v'} = C\frac{pv}{267+t}\log\frac{p'}{p}.$$

The heat released by the entry of the atmospheric air into the vacuum is therefore given by the expression when p is made very small; then $\log\frac{p'}{p}$ is very big, but the product of p with $\log\frac{p'}{p}$ is very small; for

$$p\log\frac{p'}{p} = p\log p' - p\log p = p(\log p' - \log p),$$

a quantity which converges to zero when p diminishes.

The heat released is therefore smaller the lower the pressure in the receiver, and it falls to zero when the vacuum is perfect.

We will add that the equation

$$Q = R(B-C \log p)$$

gives the law of specific heats at constant pressure and constant volume.* The first is expressed as

$$R\left(\frac{dB}{dt}-\frac{dC}{dt}\log p\right);$$

the second as

$$R\left(\frac{dB}{dt}-\frac{dC}{dt}\log p - C\frac{1}{p}\frac{dp}{dt}\right),$$

which is equal to

$$R\left(\frac{dB}{dt}-\frac{dC}{dt}\log p - \frac{C}{267+t}\right).$$

The first is obtained by differentiating Q with respect to t keeping p constant; the second keeping v constant. If equal volumes of different gases are taken at the same temperature and at the same pressure, the quantity R is the same for all and consequently it is seen that the specific heat at constant pressure exceeds the specific heat at constant volume by an amount which is the same for all gases, equal to

$$\frac{R}{267+t}\cdot C.$$

§IV

The same method applied to vapors allows us to establish a new relation between their latent heats, volumes and pressures.

We showed in the second paragraph how a liquid passing into the vapor state can serve to transmit caloric from a body maintained at a temperature T to one maintained at a lower temperature t, and how this transmission can develop motive force.

Suppose that the temperature of the body B is lower than that of

* [These are almost the only ones of Clapeyron's equations for observable quantities which are actually wrong. They imply that the specific heats of a perfect gas depend on pressure, which was consistent with some of Delaroche and Bérard's data. But he does not follow up these implications. E. M.]

the body A by an infinitesimal amount dt. Let cb (Fig. 4) represent the vapor pressure of the liquid corresponding to the temperature t

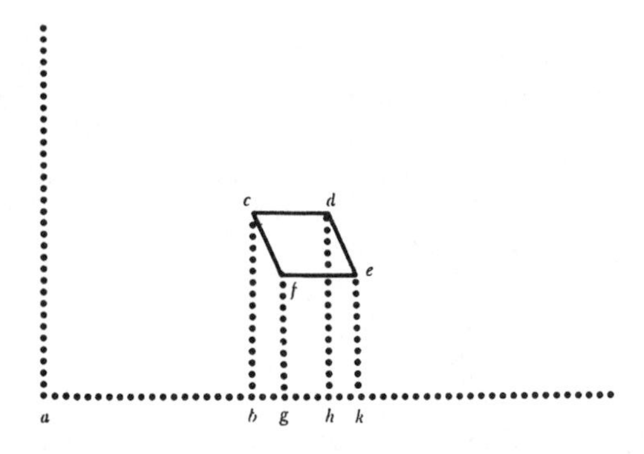

Fig. 4

of the body A, and fg that corresponding to the temperature $t–dt$ of the body B; let bh be the increase of volume due to the vapor formed by contact with the body A, hk that due to the vapor formed after removal of the body A, vapor whose formation lowered the temperature by an amount dt. Then we have seen that the quantity of action developed by the transmission of the latent caloric provided by the body A in going to the body B, is measured by the quadrilateral $cdef$. Now neglecting infinitely small quantities of the second order, this area is equal to the product of the volume cd by the differential of the pressure $dh-ek$. Calling p the vapor presure of the liquid corresponding to the temperature t, p is a function of t and we have

$$dh-ek = \frac{dp}{dt} \cdot dt.$$

cd is equal to the increase of volume which the water undergoes when passing from the liquid to the gaseous state under pressure p at the corresponding temperature. If the density of the liquid is called ϱ and that of the vapor δ, then δv is the weight and $\frac{\delta v}{\varrho}$ the volume of liquid evaporated. The increase of volume due to the formation of a volume v of vapor is therefore

$$v\left(1-\frac{\delta}{\varrho}\right).$$

Therefore the effect produced is

$$\left(1-\frac{\delta}{\varrho}\right) v \frac{dp}{dt} \cdot dt.$$

The heat which served to produce this quantity of action is the latent caloric of the volume v of vapor formed; let k be a function of t representing the latent caloric contained in unit volume of the vapor, found by experiment at temperature t at the corresponding pressure; the latent caloric of volume v is kv and the ratio of the effect produced to the heat used is expressed by

$$\frac{\left(1-\frac{\delta}{\varrho}\right)\frac{dp}{dt}\,dt}{k}.$$

We have shown that it is the greatest that it is possible to obtain, that it is independent of the nature of the liquid employed and that it is the same as that obtained through the use of the permanent gases; now we have seen that this is expressed as $\frac{dt}{C}$, C being a function of t independent of the nature of the gas, and we therefore have

$$\frac{\left(1-\frac{\delta}{\varrho}\right)\frac{dp}{dt}}{k} = \frac{1}{C}, \qquad \text{whence} \qquad k = \left(1-\frac{\delta}{\varrho}\right)\frac{dp}{dt}\,C.$$

For most vapors, the ratio $\frac{\delta}{\varrho}$ of the density of the vapor to that of the liquid is negligible compared to unity, provided that the temperature is not very high; we therefore have approximately

$$k = C\frac{dp}{dt}.$$

This equation states that *the latent caloric contained in equal volumes of the vapors of different liquids at the same temperature, under the corresponding pressures, is proportional to the coefficient of the pressure with respect to the temperature.*

It follows from this that the latent heats contained in the vapors of liquids which start to boil only at high temperatures, such as mercury for example, are very small since for these vapors the quantity $\frac{dp}{dt}$ is very small.

We will not stress the consequences which result from the equation

$$k = \left(1 - \frac{\delta}{\varrho}\right)\frac{dp}{dt}C.$$

We will limit ourselves to the remark that if, as everything leads us to believe, C and $\frac{dp}{dt}$ do not become infinite for any value of the temperature, k becomes zero when $\delta = \varrho$; that is, when the pressure becomes so great and the temperature so high that the density of the vapor becomes equal to that of the liquid, the latent heat becomes zero.

§V

Every body in nature changes its volume when its temperature or pressure changes; liquids and solids are no exceptions to this law; they can equally well serve to develop motive power from heat; it has been proposed to substitute them instead of steam in order to utilize this motive force; their use has sometimes even proved advantageous, notably when it has been necessary to develop a very great stress for a short time acting over a small distance.

In these kinds of bodies as in gases, it is to be noted that of the four quantities volume v, pressure p, temperature T and absolute quantity of heat Q, if two are determined then the other two follow; if therefore two of them are taken as independent variables, for example p and v, the other two, T and Q, can be considered as functions of them.

Direct experiments on the elasticity and thermal expansion of bodies can show how the quantities T, p and v vary together; thus, for gases, Mariotte's law concerning their elasticity and Gay-Lussac's concerning their thermal expansion lead to the equation

$$pv = R(267 + t);$$

there remains only to determine Q as a function of p and v.

There exists a relation between the functions T and Q which can be deduced from principles similar to those which we have just established. For let us increase the temperature of a body by the infinitesimal amount dT, preventing the volume from increasing by increasing the pressure; if the volume v is represented by the abscissa *ab* (Fig. 5) and the initial pressure by the ordinate *bd*, this increase in pressure can be represented by the quantity *df*, which is infinitely small, being of the same order as the increase of temperature dT which causes it.

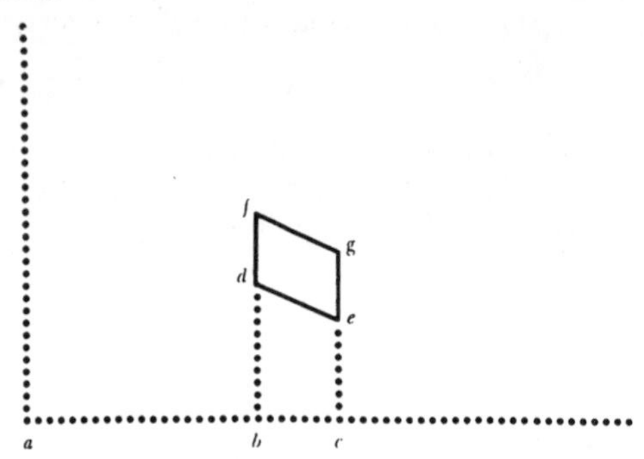

Fig. 5

We now bring up a source of heat A maintained at temperature $T+dT$ and we allow the volume v to grow by an amount *bc*, the presence of the source A maintained at temperature $T+dT$ preventing the temperature from decreasing. During the contact the quantity of heat Q which the body possesses increases by an amount dQ which is taken from the source A. Next we remove the source A and cool the given body by an amount dT keeping the volume *ac* constant. Then the pressure falls by an infinitesimal amount *ge*.

The temperature of the body is thus reduced to T which is that of the source of heat B; we bring this up and reduce the volume of the body by an amount *bc* in such a way that all the heat developed in the reduction of volume is absorbed by the body B and so that the temperature remains equal to its original value T. Since the volume v becomes again equal to that at the start of the operation, it is certain that the pressure also assumes its initial value *bd*, and at the same time it has the same absolute quantity of heat Q.

If now the four points *f*, *g*, *e* and *d* are joined by straight lines, a quadrilateral is formed whose area measures the quantity of action developed during the operation described. Now it is easy to see that *fg* and *de* are two infinitely close elements of two infinitely close curves whose equations are $T+dT=\text{const.}$, $T=\text{const.}$ They must therefore be considered parallel; since the two ordinates at the ends of the quadrilateral are also parallel, the figure is a parallelogram and its area is $bc \times df$.

Now fd is nothing more than the increment which the pressure p undergoes when T becomes $T+dT$ while the volume v remains constant. Therefore

$$df = \frac{dp}{dT}dT,$$

whence

$$fd = \frac{1}{\frac{dT}{dp}}dT.$$

And since bc is the increment of volume dv

$$fd \times bc = \frac{dv \cdot dT}{\frac{dT}{dp}}.$$

We must now determine the consumption of heat necessary for the production of this amount of mechanical action.

The first thing we did was to raise the temperature of the body by an amount dT without changing its original volume v, then when it became $v+dv$ we lowered it by the same amount dT without changing its volume $v+dv$. Now it is easy to see that this double operation can be carried out without loss of heat; for let us suppose that the interval of temperature dT is divided into a number n of new intervals $\frac{dT}{n}$, n being an infinitely large number, and that we have $n+1$ sources of heat maintained at temperatures T, $T+\frac{dT}{n}$, $T+\frac{2dT}{n}$, . . ., $T+\frac{(n-1)dT}{n}$ and $T+dT$.

To change the body on which we are operating from the temperature T to the temperature $T+dT$, we put it successively in contact with the second, third, $(n+1)$th of these sources, so that it acquires the temperature of each of them. When on the other hand the volume v of the body has increased by dv and we wish to bring it to the temperature T, we put it in contact successively with the nth, $(n-1)$th, and the first of these sources so that it acquires the temperature of each of them. The heat which was taken from each of these sources during the first part of the operation is therefore given back, since we can neglect differences of a smaller order of magnitude resulting from changes of the specific heat of the body when v and Q vary.

Each of the sources will therefore neither gain nor lose anything, with the exception however of the source whose temperature is $T+dT$, which loses the heat needed to raise the temperature of the body from $T+\frac{(n-1)dT}{n}$ to $T+dT$; and of the source maintained at temperature T which gains the heat necessary to lower the temperature of the same body from $T+\frac{dT}{n}$ to T. If n is imagined infinitely great, these quantities of heat can be neglected.

Hence it can be seen that when the body concerned has had its temperature reduced to T and has been put in contact with the source of heat B, it has gained only the heat communicated to it by the source A, since the start of the operation. As a result of the reduction of volume while in contact with the body B, it must again possess the same volume and be at the same temperature as at the beginning; the quantities Q and p must therefore have resumed their original values; it is therefore certain that all the heat taken from the source A, and nothing more than this heat, has been given to the body B.

From this it follows that the effect

$$\frac{dv \cdot dT}{\frac{dT}{dp}}$$

is due to the transmission of the heat by the body concerned during its contact with the source A, which eventually flowed into the source B.

During the contact with the source A, the temperature remained constant; whence it follows that the variations of the pressure and volume, dp and dv, are connected by the relation

$$\frac{dT}{dp}dp+\frac{dT}{dv}dv = 0.$$

These variations dp and dv produce a variation in the absolute quantity of heat Q given by

$$dQ = \frac{dQ}{dp}dp+\frac{dQ}{dv}dv = dv\left[\frac{dQ}{dv}-\frac{dQ}{dp}\frac{\left(\frac{dT}{dv}\right)}{\left(\frac{dT}{dp}\right)}\right];$$

this is the quantity of heat consumed in order to produce the effect we have calculated. The effect produced by unit quantity of heat is therefore

$$\frac{dT}{\frac{dQ}{dv}\cdot\frac{dT}{dp}-\frac{dQ}{dp}\cdot\frac{dT}{dv}}.$$

It can be demonstrated that, as for gases, this resultant effect is the greatest attainable; and since all bodies in nature can be used, in the way we have just indicated, to produce this maximum effect, this must necessarily be the same for all.

When we applied this theory particularly to gases, we called the coefficient of dT in the expression for this maximum quantity of action, $\frac{1}{C}$; we therefore have for all bodies in nature—solids, liquids or gases—the equation

$$\frac{dQ}{dv}\cdot\frac{dT}{dp}-\frac{dQ}{dp}\cdot\frac{dT}{dv} = C,$$

in which C is a function of temperature which is always the same.

For gases we have

$$T = -267+\frac{1}{R}\cdot pv,$$

whence is deduced

$$\frac{dT}{dp} = \frac{v}{R} \quad \text{and} \quad \frac{dT}{dv} = \frac{p}{R}.$$

The preceding equation, applied to gases, therefore takes the form

$$v\frac{dQ}{dv}-p\frac{dQ}{dp} = RC = F(p, v);$$

this is the equation which we have already deduced directly, and which has for integral

$$Q = R(B-C\log p);$$

the integral of the general equation

$$\frac{dQ}{dv}\cdot\frac{dT}{dp}-\frac{dQ}{dp}\cdot\frac{dT}{dv} = C$$

is of the form

$$Q = F(T)-C.\phi(p, v);$$

$F(T)$ is an arbitrary function of the temperature and $\phi(p, v)$ a particular function satisfying the equation

$$\frac{dT}{dv}\cdot\frac{d\phi}{dp}-\frac{dT}{dp}\cdot\frac{d\phi}{dv} = 1 \qquad \text{(see note at end).}$$

We shall now deduce some consequences of the general equation which we have arrived at.

We have seen in the foregoing that when a body is compressed by an amount dv, the temperature remaining constant, the heat released by the condensation is equal to

$$dQ = dv\left[\frac{dQ}{dv}-\frac{dQ}{dp}\cdot\frac{\left(\frac{dT}{dv}\right)}{\left(\frac{dT}{dp}\right)}\right];$$

and as

$$\frac{dQ}{dv}\cdot\frac{dT}{dp}-\frac{dQ}{dp}\cdot\frac{dT}{dv} = C,$$

the preceding equation can be put into the form

$$dQ = dv\cdot\frac{C}{\left(\frac{dT}{dp}\right)} = -dp\frac{C}{\left(\frac{dT}{dv}\right)}.$$

This last equation can be put in the form

$$dQ = -dp\, C\frac{dv}{dT};^*$$

$\frac{dv}{dT}$ is the differential coefficient of the volume with respect to the temperature at constant pressure.

We are therefore led to this general law which applies to all bodies in nature—solids, liquids and gases: *if the pressure supported by different bodies is increased by a small amount, quantities of heat are released which are proportional to their expansion coefficients.*

This result is the most general consequence which can be deduced from the axiom that it is absurd to imagine that motive force or heat can be created freely out of nothing.

---

* [This is a special case of $T\cdot dS=C_p dT-T\left(\frac{\partial V}{\partial T}\right)_p dp$, and it contains one of Maxwell's relations which Clapeyron deduces from the assumption that dQ is a perfect differential. E. M.]

§VI

It can be seen that the function C of temperature is of great importance because of the part which it plays in the theory of heat; it enters into the expression for the latent caloric which all bodies possess and which can be released by pressure. Unfortunately there are no experiments which allow us to determine the values of this function at all values of the temperature; for $t = 0$ it can be determined in the following way.

Dulong has shown that air or any other gas, at 0° and at 76 cm of mercury pressure, if compressed by $\frac{1}{267}$ of its volume, releases an amount of heat which is capable of raising the same volume of atmospheric air by 0.421°.

Suppose we operated on one kilogram of air occupying a volume $v=0.770$ cubic meters, under atmospheric pressure p equivalent to 10,230 kilograms per square meter; we have

$$pv = R(267+t)$$

and

$$Q = R(B-C\cdot\log p).$$

If we suddenly vary v by an infinitely small amount dv, without changing the absolute quantity of heat Q, we have

$$pdv+vdp = Rdt$$

and

$$O = R\left(\frac{dB}{dt}-\log p\cdot\frac{dC}{dt}\right)dt-RC\frac{dp}{p},$$

that is

$$\frac{dt}{C}R\left(\frac{dB}{dt}-\log p\frac{dC}{dt}\right) = R\frac{dp}{p} = \frac{R}{p}\left(\frac{Rdt-pdv}{v}\right) = \frac{Rdt-pdv}{267+t}.$$

Now since $R\left(\frac{dB}{dt}-\frac{dC}{dt}\log p\right)$ is the partial differential of Q with respect to t keeping p constant, it is nothing else than the specific heat of air at constant pressure, that is the number of units of heat necessary to raise one kilogram of air by one degree under atmospheric pressure; we therefore have

$$R\left(\frac{dB}{dt}-\frac{dC}{dt}\log p\right) = 0.267.$$

Then replacing dv by $-\frac{v}{267}$ and dt by 0.421, we find

$$\frac{1}{C} = 1.41.*$$

This is the maximum effect which can be produced by an amount of heat equal to that which would raise 1 kilogram of water at zero degrees by 1°, passing from a body maintained at 1° to a body maintained at 0°. It is expressed in kilograms raised through one meter.

Knowing the value of C corresponding to $t=0$, it is interesting to know if C increases or decreases away from this point, and how it does so. An experiment of Delaroche and Bérard on the variation of the specific heat of air with pressure allows the differential coefficient $\frac{dC}{dt}$ to be calculated.

For according to our formulae, the specific heat of air at two pressures p and p' varies as $R\frac{dC}{dt}\log\frac{p}{p'}$; equating this quantity to the difference of specific heats as deduced from the results of Delaroche and Bérard, we find

$$\frac{dC}{dt} = 0.002565$$

taking the mean of two experiments. In these experiments the air entered the calorimeter at 96.90° and left at 22.83°. 0.002565 is therefore the mean value of the differential coefficient $\frac{dC}{dt}$ between these two temperatures.

This result shows that between these limits the function C increases, but very slowly; as a result the function $\frac{1}{C}$ decreases; whence the effect produced by heat decreases at high temperatures, although very slowly.

---

* [C is in fact equal to T/J; $J=427$ kilogram-meters per kilocalorie, so that $\frac{1}{C}=1.56$ of Clapeyron's units at 273°K. The discrepancy arises mostly from the inconsistencies between Dulong's datum (which gives $C_p/C_v=1.412$ for air, implying $C_p=6.80$ cals/mol) and the direct measurements (which give $C_p=0.267$ kilocalories per kilogram, that is 7.70 cals/mol). E. M.]

The theory of vapors can give us new values of the function C at other temperatures. Let us again consider the formula

$$\frac{1}{C} = \frac{\left(1 - \frac{\delta}{\varrho}\right)\frac{dp}{dt}}{k}$$

which we deduced in paragraph IV. If we neglect the density of the vapor in comparison with that of the liquid, this reduces to

$$\frac{1}{C} = \frac{\frac{dp}{dt}}{k}.$$

It should be noted in passing that at the boiling point $\frac{dp}{dt}$ is almost the same for all vapors; C itself varies but little with temperature, so that k is itself nearly constant. This explains why physicists believed that at their boiling points, equal volumes of all vapors contain the same quantity of latent caloric; but at the same time it can be seen that this law is only approximate, since it assumes that C and $\frac{dp}{dt}$ are the same for all vapors at their boiling points.

Experiments performed by a number of physicists allow the values of k and $\frac{dp}{dt}$ to be calculated for different liquids at their boiling points; we can therefore derive the corresponding values of $\frac{1}{C}$ from them. We have thus been able to draw up the following table:

Names of liquids	Value in atmospheres of $\frac{dp}{dT}$ at the boiling point	Density of vapor at b. pt., that of air = 1	Quantity of latent heat contained in 1 kgm. of vapor	Temperature of boiling	Values of $\frac{1}{C}$
Sulfuric ether	$\frac{1}{28.12}$	2.280	90.8	35.5	1.365
Alcohol	$\frac{1}{25.19}$	1.258	207.7	78.8	1.208
Water	$\frac{1}{29.1}$	0.451	543.0	100	1.115
Oil of turpentine	$\frac{1}{30}$	3.207	76.8	156.8	1.076

These results confirm in a striking way the theory which we are expounding; they show that C increases slowly with temperature, as we have already found: we have seen that for $t=0$, $\frac{1}{C}=1.41$ so that $C = 0.7092$, this result being derived from experiments on the velocity of sound. Here we find that from experiments on steam for $t=100°$, $\frac{1}{C}=1.115$, whence $C=0.8969$; C has therefore increased by 0.187 between 0 and 100°, which gives for the mean differential coefficient

$$\frac{dC}{dt} = 0.00187$$

between these limits.

The mean of two experiments by Delaroche and Bérard gives us for the mean value of $\frac{dC}{dt}$ between 22.83° and 96.90°, the quantity

$$\frac{dC}{dt} = 0.002565.$$

These two results hardly differ from one another, and their divergence can be well understood when we consider the number and variety of the experiments from which the data are taken.

There exists another means of calculating $\frac{1}{C}$ approximately, between wide limits of temperature; it is necessary to assume that the quantity of caloric contained in a given weight of steam is the same at all temperatures under the corresponding pressures,* and further that the laws concerning the compression and expansion of permanent gases apply equally well to vapors; if these are adopted, though they are only approximate, then in the formula

$$\frac{1}{C} = \frac{\frac{dp}{dt}}{k}$$

k can be expressed as a function of t; $\frac{dp}{dt}$ can be derived in the range 0° to 100° from old experiments by various physicists, and from 100° to 224° from recent experiments by Arago and Dulong.

* [This is Watt's law, that the enthalpy of saturated steam is constant. With the data at 100°C as standard, it gives the latent heats at 0°C and 157°C correct to about 6%. E. M.]

Then it is found for

	We have already found for the same values of t these values of $1/C$: Values of $\frac{1}{C}$
$t=0, \frac{1}{C}=1.586$	1.410.
$t=35.5, \frac{1}{C}=1.292$	1.365.
$t=78.8, \frac{1}{C}=1.142$	1.208.
$t=100, \frac{1}{C}=1.102$	1.115.
$t=156.8, \frac{1}{C}=1.072$	1.076.

This last column, deduced from experiments on sound, on the vapors of sulfuric ether, alcohol, water and turpentine oil, is in satisfactory agreement with the first.

These remarkable numerical coincidences, obtained from a great number of diverse data taken from very different phenomena, seem to add a great deal to the evidence for our theory.

§VII

The function C is, as we have seen, of great importance: it is the common link between the phenomena caused by heat in solid bodies, liquids and gases; it would be desirable for very exact experiments, such as researches on the propagation of sound at different temperatures, to be done to allow this function to be found with the desired accuracy; this would determine several other important things in the theory of heat, concerning which experiments have only led to poor approximations or have as yet told us nothing. Among these we may include the heat released by the compression of solids or liquids; the theory which we have given allows us to determine it numerically for all values of temperature for which the function C is known with sufficient accuracy, that is to say between $t=0°$ and $224°$.

We have seen that the heat set free by an increase of pressure dp is equal to the thermal expansion of the body under experiment, multiplied by C. For air at zero degrees, the quantity of heat released can be deduced at once from experiments on the velocity of sound, in the following way.

Dulong has shown that a compression by $\frac{1}{267}$ raises the temperature of a volume of air at zero degrees by 0.421°. Now the 0.267 units of heat necessary to raise one kilogram of air at zero degrees under constant pressure by 1°, are composed of the heat needed to maintain the temperature of the gas at zero degrees while dilated by $\frac{1}{267}$ of its volume, plus the heat needed for raising it by 1° at constant volume after expansion; this last is equal to $\frac{1}{0.421}$ of the first: their sum is therefore equal to the first multiplied by $1+\frac{1}{0.421}$; therefore the latter, i.e. the heat necessary for maintaining the temperature of one kilogram of air, dilated by $\frac{1}{267}$ of its volume, at zero degrees, is equal to $(0.267):(1+\frac{1}{0.421})$, that is, to 0.07911.

The same result can be arrived at using the formula

$$Q = R(B-C \log p),$$

whence

$$dQ = RC\frac{dv}{v},$$

putting $C=\frac{1}{1.410}$ and noting that a diminution of volume of $\frac{1}{267}$ corresponds to an increase of pressure of $\frac{1}{267}$ of an atmosphere.

Knowing the heat set free by the compression of gases, we can find that released by a similar pressure acting on any other body, for example iron, by writing the proportion: 0.07911, the heat released by a volume of air equal to 0.77 cubic meters acted on by an increment of pressure equal to $\frac{1}{267}$ of an atmosphere, is, to that released from an equal volume of iron in the same circumstances, as 0.00375, the cubical thermal expansion of air, is to 0.00003663, the cubical thermal expansion of iron. For the second term of the proportion we find the number 0.0007718. Now a volume of 0.77 cubic meters of iron weighs 5996 kilograms: the heat set free by one kilogram is therefore $\frac{0.0007718}{5996}$; for a pressure of 1 atmosphere it is 267 times greater, or equal to 0.00003436; dividing this number by the specific heat of iron with respect to that of water, the amount by which the temperature of iron is raised by one atmosphere is found; it is seen to be too small to be observed by our thermometers.

§ VIII

We will not dwell further on the consequences for the theory of heat which come from the results announced in this Memoir: we believe it is useful merely to add a few words on the use of heat as a motive force. S. Carnot, in the work already cited, seems to us to have laid the real foundations of this important part of practical mechanics.

High- or low-pressure machines without expansion use the *vis viva* which the caloric contained in the steam can develop in its passage from the temperature of the boiler to that of the condenser; in high-pressure machines without condensers, everything takes place as if they were provided with condensers at 100°. In these, therefore, we only utilize the passage of the latent heat contained in the vapor from the temperature of the boiler to 100°. The perceptible caloric of the vapor is entirely lost in all machines without expansion.

It is utilized in part by machines with expansion in which the temperature of the vapor is allowed to fall; the outer cylinder, which in Woolf double-cylinder machines is designed to keep the temperature of the steam constant—which is very useful for reducing the limits between which the motive force acting on the piston fluctuates—has only an undesirable influence on the effect produced in terms of fuel consumption.

To utilize completely the motive force available, it would be necessary to push the expansion till the temperature of the steam fell to that of the condenser; but practical considerations, deriving from the way in which the motive force of fire is utilized in industry, hinder the attainment of this limit.

We have shown elsewhere that the use of gases or any liquid other than water, between the same limits of temperature, would not improve on results already obtained; but it does emerge from the foregoing considerations that since the temperature of fire is more than 1000° or 2000° higher than that in a boiler, there is an enormous loss of *vis viva* in the passage of the heat from the furnace to the boiler. It is solely through the use of caloric at high temperatures, and through the discovery of media suitable for extracting its motive power, that important improvements in the art of utilizing the motive power of heat may be expected.

NOTE

The integral of the general equation

$$\frac{dQ}{dv}\cdot\frac{dT}{dp}-\frac{dQ}{dp}\cdot\frac{dT}{dv}=C$$

is, as we have seen,

$$Q=F(T)-C.\phi(p,v). \qquad (1)$$

$F(T)$ is an arbitrary function of the temperature T, which can vary from one body to another; C is a function of temperature which is the same for all bodies in nature, and $\phi(p, v)$ is a particular function of p and v satisfying the equation

$$\frac{dT}{dv}\cdot\frac{d\phi}{dp}-\frac{dT}{dp}\cdot\frac{d\phi}{dv}=1. \qquad (2)$$

This function ϕ can be found in the following way. Let

$$\phi=\int\frac{dp}{\frac{dT}{dv}}+\phi';$$

then substituting in equation (2) we find

$$\frac{dT}{dv}\cdot\frac{d\phi'}{dp}-\frac{dT}{dp}\cdot\frac{d\phi'}{dv}=\frac{dT}{dp}\cdot\frac{d}{dv}\int\frac{dp}{\frac{dT}{dv}}.$$

This equation can be satisfied by putting

$$\phi'=\int dp\frac{\frac{dT}{dp}}{\frac{dT}{dv}}\frac{d}{dv}\int\frac{dp}{\frac{dT}{dv}}+\phi'';$$

ϕ'' satisfies the equation

$$\frac{dT}{dv}\cdot\frac{d\phi''}{dp}-\frac{dT}{dp}\cdot\frac{d\phi''}{dv}=\frac{dT}{dp}\cdot\frac{d}{dv}\int dp\frac{\frac{dT}{dp}}{\frac{dT}{dv}}\cdot\frac{d}{dv}\int\frac{dp}{\frac{dT}{dv}}.$$

We also have

$$\phi''=\int dp\frac{\frac{dT}{dp}}{\frac{dT}{dv}}\cdot\frac{d}{dv}\int dp\frac{\frac{dT}{dp}}{\frac{dT}{dv}}\frac{d}{dv}\int\frac{dp}{\frac{dT}{dv}}+\phi'''.$$

Thus it can be seen that $\phi(p, v)$ is given by a series of terms, each one of which is found from the preceding one by differentiating it with respect to v, multiplying by the ratio $\dfrac{\frac{dT}{dp}}{\frac{dT}{dv}}$ and integrating the result with respect to p. Since the first term of this series is $\displaystyle\int \frac{dp}{\frac{dT}{dv}}$, it is seen that the value of ϕ can be easily found; substituting this value in equation (1), we have the following expression for the general integral of the partial differential equation

$$\frac{dQ}{dv}\cdot\frac{dT}{dp}-\frac{dQ}{dp}\cdot\frac{dT}{dv} = C,$$

the formula

$$Q = F(t) - C \int \frac{dp}{\frac{dT}{dv}}$$
$$+ \int dp \frac{\frac{dT}{dp}}{\frac{dT}{dv}}\cdot\frac{d}{dv}\int \frac{dp}{\frac{dT}{dv}}$$
$$+ \int dp \frac{\frac{dT}{dp}}{\frac{dT}{dv}}\cdot\frac{d}{dv}\int dp \frac{\frac{dT}{dp}}{\frac{dT}{dv}}\cdot\frac{d}{dv}\int \frac{dp}{\frac{dT}{dv}}$$
$$+ \cdots$$

This equation supplies the law of specific heats and of caloric released by the variations of volume and pressure of all bodies in nature, when one knows the relation which exists between temperature, volume and pressure.

ON THE MOTIVE POWER OF HEAT, AND ON THE LAWS WHICH CAN BE DEDUCED FROM IT FOR THE THEORY OF HEAT

BY R. CLAUSIUS

(Poggendorff's *Annalen der Physik,* LXXIX, [1850] 368, 500)

Translated by
W. F. MAGIE

Rudolf Clausius.

On the Motive Power of Heat, and on the Laws which can be Deduced from it for the Theory of Heat

SINCE heat was first used as a motive power in the steam-engine, thereby suggesting from practice that a certain quantity of work may be treated as equivalent to the heat needed to produce it, it was natural to assume also in theory a definite relation between a quantity of heat and the work which in any possible way can be produced by it, and to use this relation in drawing conclusions about the nature and the laws of heat itself. In fact, several fruitful investigations of this sort have already been made; yet I think that the subject is not yet exhausted, but on the other hand deserves the earnest attention of physicists, partly because serious objections can be raised to the conclusions that have already been reached, partly because other conclusions, which may readily be drawn and which will essentially contribute to the establishment and completion of the theory of heat, still remain entirely unnoticed or have not yet been stated with sufficient definiteness.

The most important of the researches here referred to was that of S. Carnot,* and the ideas of this author were afterwards given analytical form in a very skilful way by Clapeyron.† Carnot showed that whenever work is done by heat and no permanent change occurs in the condition of the working body, a certain quantity of heat passes from a hotter to a colder body. In the steam-engine, for example, by means of the steam which is developed in the boiler and precipitated in the condenser, heat is transferred from the grate to the condenser. This *transfer* he considered as

* *Réflexions sur la puissance motrice du feu, et sur les machines propres à développer cette puissance, par S. Carnot. Paris*, 1824. I have not been able to obtain a copy of this book, and am acquainted with it only through the work of Clapeyron and Thomson, from the latter of whom are quoted the extracts afterwards given.

† *Journ. de l'École Polytechnique*, vol. xix (1834), and Pogg. *Ann.*, vol. lix.

the heat change, corresponding to the work done. He says expressly that no heat is lost in the process, but that the *quantity of heat* remains unchanged, and adds: "This fact is not doubted; it was assumed at first without investigation, and then established in many cases by calorimetric measurements. To deny it would overthrow the whole theory of heat, of which it is the foundation." I am not aware, however, that it has been sufficiently proved by experiment that no loss of heat occurs when work is done; it may, perhaps, on the contrary, be asserted with more correctness that even if such a loss has not been proved directly, it has yet been shown by other facts to be not only admissible, but even highly probable. It it be assumed that heat, like a substance, cannot diminish in quantity, it must also be assumed that it cannot increase. It is, however, almost impossible to explain the heat produced by friction except as an increase in the quantity of heat. The careful investigations of Joule, in which heat is produced in several different ways by the application of mechanical work, have almost certainly proved not only the possibility of increasing the quantity of heat in any circumstances but also the law that the quantity of heat developed is proportional to the work expended in the operation. To this it must be added that other facts have lately become known which support the view, that heat is not a substance, but consists in a motion of the least parts of bodies. If this view is correct, it is admissible to apply to heat the general mechanical principle that a motion may be transformed into work, and in such a manner that the loss of *vis viva* is proportional to the work accomplished.

These facts, with which Carnot also was well acquainted, and the importance of which he has expressly recognized, almost compel us to accept the equivalence between heat and work, on the modified hypothesis that the accomplishment of work requires not merely a change in the distribution of heat, but also an actual consumption of heat, and that, conversely, heat can be developed again by the expenditure of work.

In a memoir recently published by Holtzmann,* it seems at first as if the author intended to consider the matter from this latter point of view. He says (p. 7): "The action of the heat supplied to the gas is either an elevation of temperature, in conjunction with an increase in its elasticity, or mechanical work, or

* *Ueber die Wärme und Elasticität der Gase und Dämpfe*, von C. Holtzmann, Mannheim, 1845; also Pogg. *Ann.*, vol. 72a.

a combination of both, and the mechanical work is the equivalent of the elevation of temperature. The heat can only be measured by its effects; of the two effects mentioned the mechanical work is the best adapted for this purpose, and it will accordingly be so used in what follows. I call the unit of heat the heat which by its entrance into a gas can do the mechanical work *a*—that is, to use definite units, which can lift *a* kilograms through 1 meter." Later (p. 12) he also calculates the numerical value of the constant *a* in the same way as Mayer had already done,* and obtains a number which corresponds with the heat equivalent obtained by Joule in other entirely different ways. In the further extension of his theory, however, in particular in the development of the equations from which his conclusions are drawn, he proceeds exactly as Clapeyron did, so that in this part of his work he tacitly assumes that the quantity of heat is constant.

The difference between the two methods of treatment has been much more clearly grasped by W. Thomson, who has extended Carnot's discussion by the use of the recent observations of Regnault on the tension and latent heat of water vapor.† He speaks of the obstacles which lie in the way of the unrestricted assumption of Carnot's theory, calling special attention to the researches of Joule, and also raises a fundamental objection which may be made against it. Though it may be true in any case of the production of work, when the working body has returned to the same condition as at first, that heat passes from a warmer to a colder body, yet on the other hand it is not generally true that whenever heat is transferred work is done. Heat can be transferred by simple conduction, and in all such cases, if the mere transfer of heat were the true equivalent of work, there would be a loss of working power in Nature, which is hardly conceivable. Nevertheless, he concludes that in the present state of the science the principle adopted by Carnot is still to be taken as the most probable basis for an investigation of the motive power of heat, saying: "If we abandon this principle, we meet with innumerable other difficulties—insuperable without further experimental investigation—and an entire reconstruction of the theory of heat from its foundation."‡

I believe that we should not be daunted by these difficulties,

---

* *Ann. der Chem. und Pharm.* of Wöhler and Liebig, vol. xlii., p. 239.

† *Transactions of the Royal Society of Edinburgh*, vol. xvi.

‡ *Math. and Phys. Papers*, vol. i, p. 119, note.

but rather should familiarize ourselves as much as possible with the consequences of the idea that heat is a motion, since it is only in this way that we can obtain the means wherewith to confirm or to disprove it. Then, too, I do not think the difficulties are so serious as Thomson does, since even though we must make some changes in the usual form of presentation, yet I can find no contradiction with any proved facts. It is not at all necessary to discard Carnot's theory entirely, a step which we certainly would find it hard to take, since it has to some extent been conspicuously verified by experience. A careful examination shows that the new method does not stand in contradiction to the essential principle of Carnot, but only to the subsidiary statement *that no heat is lost*, since in the production of work it may very well be the case that at the same time a certain quantity of heat is consumed and another quantity transferred from a hotter to a colder body, and both quantities of heat stand in a definite relation to the work that is done. This will appear more plainly in the sequel, and it will there be shown that the consequences drawn from the two assumptions are not only consistent with one another, but are even mutually confirmatory.

I. CONSEQUENCES OF THE PRINCIPLE OF THE EQUIVALENCE OF HEAT AND WORK

We shall not consider here the kind of motion which can be conceived of as taking place within bodies, further than to assume in general that the particles of bodies are in motion, and that their heat is the measure of their *vis viva*, or rather still more generally, we shall only lay down a principle conditioned by that assumption as a fundamental principle, in the words: In all cases in which work is produced by the agency of heat, a quantity of heat is consumed which is proportional to the work done; and, conversely, by the expenditure of an equal quantity of work an equal quantity of heat is produced.

Before we proceed to the mathematical treatment of this principle, some immediate consequences may be premised which affect our whole method of treatment, and which may be understood without the more definite demonstration which will be given them later by our calculations.

It is common to speak of the *total heat* of bodies, especially of gases and vapors, by which term is understood the sum of the free and latent heat, and to assume that this is a quantity dependent only on the actual condition of the body considered, so that, if all its other physical properties, its temperature, its density, etc., are known, the total heat contained in it is completely determined. This assumption, however, is no longer admissible if our principle is adopted. Suppose that we are given a body in a definite state—for example, a quantity of gas with the temperature t_0 and the volume v_0—and that we subject it to various changes of temperature and volume, which are such, however, as to bring it at last to its original state again. According to the common assumption, its total heat will again be the same as at first, from which it follows that if during one part of its changes heat is communicated to it from without, the same quantity of heat must be given up by it in the other part of its changes. Now with every change of volume a certain amount of work must be done by the gas or upon it, since by its expansion it overcomes an external pressure, and since its compression can be brought about only by an exertion of external pressure. If, therefore, among the changes to which it has been subjected there are changes of volume, work must be done upon it and by it. It is not necessary, however, that at the end of the operation, when it is again brought to its original state, the work done by it shall on the whole equal that done upon it, so that the two quantities of work shall counterbalance each other. There may be an excess of one or the other of these quantities of work, since the compression may take place at a higher or lower temperature than the expansion, as will be more definitely shown later on. To this excess of work done by the gas or upon it there must correspond, by our principle, a proportional excess of heat consumed or produced, and the gas cannot give up to the surrounding medium the same amount of heat as it receives.

The same contradiction to the ordinary assumption about the *total heat* may be presented in another way. If the gas at t_0 and v_0 is brought to the higher temperature t_1 and the larger volume v_1, the quantity of heat which must be imparted to it is, on that assumption, independent of the way in which the change is brought about; from our principle, however, it is different, according as the gas is first heated while its volume, v_0, is constant, and then allowed to expand at the constant temperature t_1, or is first expanded at the constant temperature t_0, and then heated, or as the expansion and heating are interchanged in any other way or even occur together, since in all these cases the work done by the gas is different.

In the same way, if a quantity of water at the temperature t_0 is changed into vapor at the temperature t_1 and of the volume v_1, it will make a difference in the amount of heat needed if the water as such is first heated to t_1 and then evaporated, or if it is evaporated at t_0 and the vapor then brought to the required volume and temperature, v_1 and t_1, or finally if the evaporation occurs at any intermediate temperature.

From these considerations and from the immediate application of the principle, it may easily be seen what conception must be formed of *latent* heat. Using again the example already employed, we distinguish in the quantity of heat which must be imparted to the water during its changes the *free* and *latent* heat. Of these, however, we may consider only the former as really present in the vapor that has been formed. The latter is not merely, as its name implies, *concealed* from our perception, but it is *nowhere present*; it is *consumed* during the changes in doing work.

In the heat consumed we must still introduce a distinction—that is to say, the work done is of two kinds. First, there is a certain amount of work done in overcoming the mutual attractions of the particles of the water, and in separating them to such a distance from one another that they are in the state of vapor. Secondly, the vapor during its evolution must push back an external pressure in order to make room for itself. The former work we shall call the *internal*, the latter, the *external* work, and shall partition the latent heat accordingly.

It can make no difference with respect to the *internal* work whether the evaporation goes on at t_0 or at t_1, or at any intermediate temperature, since we must consider the attractive force of the particles, which is to be overcome, as invariable.*

The *external* work, on the other hand, is regulated by the pressure as dependent on the temperature. Of course the same is true in general as in this special example, and therefore if it was said above that the quantity of heat which must be imparted to a body, to

* It cannot be raised, as an objection to this statement, that water at t_1 has less cohesion than at t_0, and that therefore less work would be needed to overcome it. For a certain amount of work is used in diminishing the cohesion, which is done while the water as such is heated, and this must be reckoned in with that done during the evaporation. It follows at once that only a part of the heat, which the water takes up from without while it is being heated, is to be considered as free heat, while the remainder is used in diminishing the cohesion. This view is also consistent with the circumstance that water has so much greater a specific heat than ice, and probably also than its vapor.

bring it from one condition to another, depended not merely on its initial and final conditions, but also on the way in which the change takes place, this statement refers only to that part of the *latent* heat which corresponds to the *external* work. The other part of the *latent* heat, as also the *free* heat, are independent of the way in which the changes take place.

If now the vapor at t_1 and v_1 is again transformed into water, work will thereby be *expended*, since the particles again yield to their attractions and approach each other, and the external pressure again advances. Corresponding to this, heat must be *produced*, and the so-called liberated heat which appears during the operation does not merely come out of concealment but is actually made new. The heat produced in this reversed operation need not be equal to that used in the direct one, but that part which corresponds to the *external* work may be greater or less according to circumstances.

We shall now turn to the mathematical discussion of the subject, in which we shall restrict ourselves to the consideration of the *permanent gases* and of *vapors at their maximum density*, since these cases, in consequence of the extensive knowledge we have of them, are most easily submitted to calculation, and besides that are the most interesting.

Let there be given a certain quantity, say a unit of weight, of a *permanent gas*. To determine its present condition, three magnitudes must be known: the pressure upon it, its volume, and its temperature. These magnitudes are in a mutual relationship, which is expressed by the combined laws of Mariotte and Gay-Lussac,* and may be represented by the equation:

$$\text{(I.)} \qquad pv = R\,(a+t),$$

where p, v, and t represent the pressure, volume, and temperature of the gas in its present condition, a is a constant, the same for all gases, and R is also a constant, which in its complete form is $\frac{p_0 v_0}{a+t_0}$, if p_0, v_0, and t_0 are the corresponding values of the three magnitudes already mentioned for any other condition of the gas. This last constant is in so far different for the different gases that it is inversely proportional to their specific gravities.

It is true that Regnault has lately shown by a very careful

* This law will hereafter, for brevity, be called the M. and G. law, and Mariotte's law will be called the M. law.

investigation that this law is not strictly accurate, yet the departures from it are in the case of the permanent gases very small, and only become of consequence in the case of those gases which can be condensed into liquids. From this it seems to follow that the law holds with greater accuracy the more removed the gas is from its condensation point with respect to pressure and temperature. We may therefore, while the accuracy of the law for the permanent gases in their ordinary condition is so great that it can be treated as complete in most investigations, think of a limiting condition for each gas, in which the accuracy of the law is actually complete. We shall, in what follows, when we treat the permanent gases as such, assume this ideal condition.

According to the concordant investigations of Regnault and Magnus, the value of $\frac{1}{a}$ for atmospheric air is equal to 0.003665, if the temperature is reckoned in centigrade degrees from the freezing-point. Since, however, as has been mentioned, the gases do not follow the M. and G. law exactly, the same value of $\frac{1}{a}$ will not always be obtained, if the measurements are made in different circumstances. The number here given holds for the case when air is taken at 0° under the pressure of *one* atmosphere, and heated to 100° at constant volume, and the increase of its expansive force observed. If, on the other hand, the pressure is kept constant, and the increase of its volume observed, the somewhat greater number 0.003670 is obtained. Further, the numbers increase if the experiment is tried under a pressure higher than the atmospheric pressure, while they diminish somewhat for lower pressures. It is not therefore possible to decide with certainty on the number which should be adopted for the gas in the ideal condition in which naturally all differences must disappear; yet the number 0.003665 will surely not be far from the truth, especially since this number very nearly obtains in the case of hydrogen, which probably approaches the most nearly of all the gases the ideal condition, and for which the changes are in the opposite sense to those of the other gases. If we therefore adopt this value of $\frac{1}{a}$ we obtain

$$a = 273.$$

In consequence of equation (I.) we can treat any one of the three magnitudes p, v, and t—for example, p—as a function of the two others, v and t. These latter then are the independent variables by which the condition of the gas is fixed. We shall now seek to determine how the magnitudes which relate to the quantities of heat depend on these two variables.

If any body changes its volume, mechanical work will in general be either produced or expended. It is, however, in most cases impossible to determine this exactly, since besides the *external* work there is generally an unknown amount of *internal* work done. To avoid this difficulty, Carnot employed the ingenious method already referred to of allowing the body to undergo its various changes in succession, which are so arranged that it returns at last exactly to its original condition. In this case, if *internal* work is done in some of the changes, it is exactly compensated for in the others, and we may be sure that the *external* work, which remains over after the changes are completed, is all the work that has been done. Clapeyron has represented this process graphically in a very clear way, and we shall follow his presentation now for the permanent gases, with a slight alteration rendered necessary by our principle.

In the figure, let the abscissa oe represent the volume and the ordinate ea the pressure on a unit weight of gas, in a condition in which its temperature $= t$. We assume that the gas is contained in an extensible envelope, which, however, cannot exchange heat with it. If, now, it is allowed to expand in this envelope, its temperature would fall if no heat were imparted to it. To avoid this, let it be put in contact, during its expansion, with a body, A, which is kept at the constant temperature t, and which imparts just so much heat to the gas that its temperature also remains equal to t. During this expansion at constant temperature, its pressure diminishes according to the M. law, and may be represented by the ordinate of a curve, ab, which is a portion of an equilateral hyperbola. When the volume of the gas has increased in this way from oe to of, the body A is removed, and the expansion is allowed to continue without the introduction of more heat. The temperature will then fall, and the pressure diminish more rapidly than before. The law which is followed in this part of the process may be represented by the curve bc. When the volume of the gas has increased in this way from of to og, and its temperature has fallen from t to τ, we begin to compress it, in order to restore it again to its original volume oe. If it were left to itself its temperature would again rise. This, however, we do not permit, but bring it in contact with a body B, at the constant temperature τ, to which it at once gives up the heat that is produced, so that it keeps the temperature τ; and while it is in contact with this body we compress it so far (by the amount *gh*) that the remaining compression *he* is exactly sufficient to raise its temperature from τ to t, if during this last compression

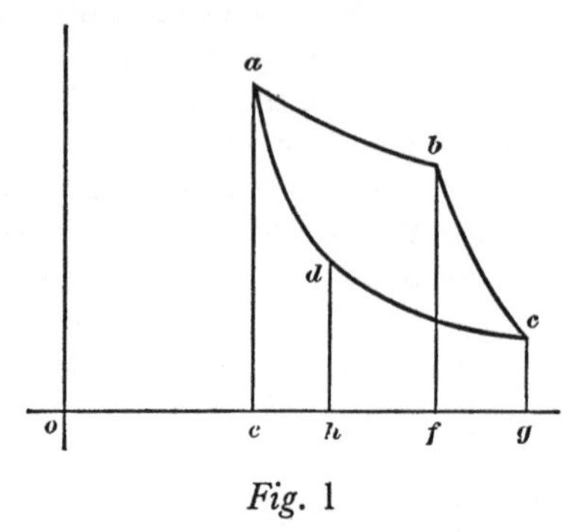

Fig. 1

it gives up no heat. During the former compression the pressure increases according to the M. law, and is represented by the portion *cd* of an equilateral hyperbola. During the latter, on the other hand, the increase is more rapid and is represented by the curve *da*. This curve must end exactly at *a*, for since at the end of the operation the volume and temperature have again their original values, the same must be true of the pressure also, which is a function of them both. The gas is therefore in the same condition again as it was at the beginning.

Now, to determine the work produced by these changes, for the reasons already given, we need to direct our attention only to the *external* work. During the expansion the gas *does* work, which is determined by the integral of the product of the differential of the volume into the corresponding pressure, and is therefore represented geometrically by the quadrilaterals *eabf* and *fbcg*. During the compression, on the other hand, work is *expended*, which is represented similarly by the quadrilaterals *gcdh* and *hdae*. The excess of the former quantity of work over the latter is to be looked on as the whole work produced during the changes, and this is represented by the quadrilateral *abcd*.

If the process above described is carried out in the reverse order, the same magnitude, *abcd*, is obtained as the excess of the work *expended* over the work *done*.

In order to make an analytical application of the method just described, we will assume that all the changes which the gas undergoes are *infinitely small*. We may then treat the curves obtained as straight lines, as they are represented in the accompanying figure. We may also, in determining the area of the quadrilateral *abcd*, consider it a parallelogram, since the error arising therefrom can only be a quantity of the *third* order, while the area itself is a quantity of the *second* order. On this assumption, as may easily be seen, the area may be represented by the product $ef \cdot bk$, if k is the point in which the ordinate bf cuts the lower side of the quadrilateral. The magnitude bk is the increase of the pressure, while

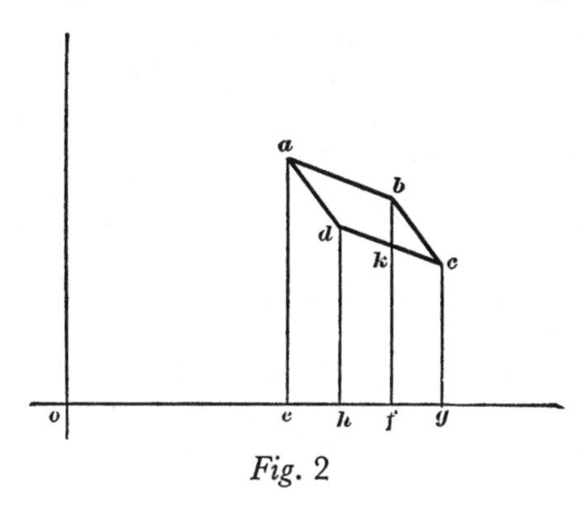

Fig. 2

the gas at the constant volume of has its temperature raised from τ to t—that is, by the differential $t-\tau=dt$. This magnitude may be at once expressed by the aid of equation (I.) in terms of v and t, and is

$$dp = \frac{Rdt}{v}.$$

If, further, we denote the increase of volume ef by dv, we obtain the area of the quadrilateral, and so, also,

$$\text{(1)} \qquad \textit{The work done} = \frac{Rdvdt}{v}.$$

We must now determine the heat consumed in these changes. The quantity of heat which must be communicated to a gas, while it is brought from any former condition in a definite way to that condition in which its volume $=v$ and its temperature $=t$, may be called Q, and the changes of volume in the above process, which must here be considered separately, may be represented as follows:

ef by dv, *hg* by $d'v$, *eh* by δv, and *fg* by $\delta' v$. During an expansion from the volume $oe=v$ to the volume $of=v+dv$ at the constant temperature t, the gas must receive the quantity of heat

$$\left(\frac{dQ}{dv}\right) dv,$$

and correspondingly, during an expansion from $oh=v+\delta v$ to $og=v+\delta v+d'v$ at the temperature $t-dt$, the quantity of heat,

$$\left[\left(\frac{dQ}{dv}\right)+\frac{d}{dv}\left(\frac{dQ}{dv}\right)\delta v-\frac{d}{dt}\left(\frac{dQ}{dv}\right)dt\right] d'v.$$

In the case before us this latter quantity must be taken as negative in the calculation, because the real process was a compression instead of the expansion assumed. During the expansion from *of* to *og* and the compression from *oh* to *oe*, the gas has neither gained nor lost heat, and hence the quantity of heat which the gas has received in excess of that which it has given up—that is, the *heat consumed*

$$(2) \qquad =\left(\frac{dQ}{dv}\right)dv-\left[\left(\frac{dQ}{dv}\right)+\frac{d}{dv}\left(\frac{dQ}{dv}\right)\delta v-\frac{d}{dt}\left(\frac{dQ}{dv}\right)dt\right] d'v.$$

The magnitudes δv and $d'v$ must be eliminated from this expression. For this purpose we have first, immediately from the inspection of the figure, the following equation:

$$dv+\delta' v = \delta v+d'v.$$

From the condition that during the compression from *oh* to *oe*, and therefore also conversely during an expansion from *oe* to *oh* occurring under the same conditions, and similarly during the expansion from *of* to *og*, both of which occasion a fall of temperature by the amount dt, the gas neither receives nor gives up heat, we obtain the equations

$$\left(\frac{dQ}{dv}\right)\delta v-\left(\frac{dQ}{dt}\right)dt = 0,$$

$$\left[\left(\frac{dQ}{dv}\right)+\frac{d}{dv}\left(\frac{dQ}{dv}\right)dv\right]\delta' v-\left[\left(\frac{dQ}{dt}\right)+\frac{d}{dv}\left(\frac{dQ}{dt}\right)dv\right]dt = 0.$$

Eliminating from these three equations and equation (2) the three magnitudes $d'v$, δv, and $\delta' v$, and also neglecting in the development those terms which, in respect of the differentials, are of a higher order than the second, we obtain

$$(3) \qquad \textit{The heat consumed} = \left[\frac{d}{dt}\left(\frac{dQ}{dv}\right)-\frac{d}{dv}\left(\frac{dQ}{dt}\right)\right] dvdt.$$

If we now return to our principle, that to produce a certain amount of work the expenditure of a proportional quantity of heat is necessary, we can establish the formula

$$\text{(4)} \qquad \frac{\textit{The heat consumed}}{\textit{The work done}} = A,$$

where A *is a constant, which denotes the heat equivalent for the unit of work.* The expressions (1) and (3) substituted in this equation give

$$\frac{\left[\frac{d}{dt}\left(\frac{dQ}{dv}\right) - \frac{d}{dv}\left(\frac{dQ}{dt}\right)\right] dvdt}{\frac{R \cdot dvdt}{v}} = A,$$

or

$$\text{(II.)} \qquad \frac{d}{dt}\left(\frac{dQ}{dv}\right) - \frac{d}{dv}\left(\frac{dQ}{dt}\right) = \frac{AR}{v}.$$

We may consider this equation as the analytical expression of our fundamental principle applied to the case of permanent gases. It shows that Q cannot be a function of v and t, if these variables are independent of each other. For if it were, then by the well-known law of the differential calculus, that if a function of two variables is differentiated with respect to both of them, the order of differentiation is indifferent, the right-hand side of the equation should be equal to zero.

The equation may also be brought into the form of a *complete* differential equation,

$$\text{(II.}a\text{)} \qquad dQ = dU + \mathrm{A} \cdot \mathrm{R} \frac{a+t}{v} dv,$$

in which U is an arbitrary function of v and t. This differential equation is naturally not integrable, but becomes so only if a second relation is given between the variables, by which t may be treated as a function of v. The reason for this is found in the last term, and this corresponds exactly to the *external* work done during the change, since the differential of this work is pdv, from which we obtain

$$\frac{R(a+t)}{v} dv,$$

if we eliminate p by means of (I.).

We have thus obtained from equation (II.*a*) what was introduced before as an immediate consequence of our principle, that the total amount of heat received by the gas during a change of volume and temperature can be separated into two parts, one of which, U, which comprises the *free* heat that has entered and the heat *consumed* in doing *internal* work, if any such work has been done, has the properties which are commonly assigned to the total heat, of being a function of v and t, and of being therefore fully determined by the initial and final conditions of the gas, between which the transformation has taken place; while the other part, which comprises the heat *consumed* in doing *external* work, is dependent not only on the terminal conditions, but on the whole course of the changes between these conditions.

Before we undertake to prepare this equation for further conclusions, we shall develop the analytical expression of our fundamental principle for the case of vapors at their maximum density.

In this case we have no right to apply the M. and G. law, and so must restrict ourselves to the principle alone. In order to obtain an equation from it, we again use the method given by Carnot and graphically presented by Clapeyron, with a slight modification. Consider a liquid contained in a vessel impenetrable by heat, of which, however, only a part is filled by the liquid, while the rest is left free for the vapor, which is at the maximum density corresponding to its temperature, t. The total volume of both liquid and vapor is represented in the accompanying figure by the abscissa *oe*, and the pressure of the vapor by the ordinate *ea*. Let the vessel now yield to the pressure and enlarge in volume while the liquid and vapor are in contact with a body A, at the constant temperature t. As the volume increases, more liquid evaporates, but the heat which thus becomes latent is supplied from the body A, so that the temperature, and so also the pressure, of the vapor remain unchanged. If in this way the total volume is increased from *oe* to *of*, an amount of external work is done which is represented by the rectangle *eabf*. Now remove the body A and let the vessel increase in volume still further, while heat can neither enter nor leave it. In this processs the vapor already present will expand, and also new vapor will be produced, and in consequence the temperature will fall and the pressure diminish. Let this process go on until the temperature has changed from t or τ, and the volume has become *og*. If the fall of pressure during this expansion is represented by the curve *bc*, the external work done in the process $= fbcg$.

Now diminish the volume of the vessel, in order to bring the liquid with its vapor back to its original total volume, *oe*; and let this compression take place, in part, in contact with the body *B* at the temperature τ, into which body all the heat set free by the

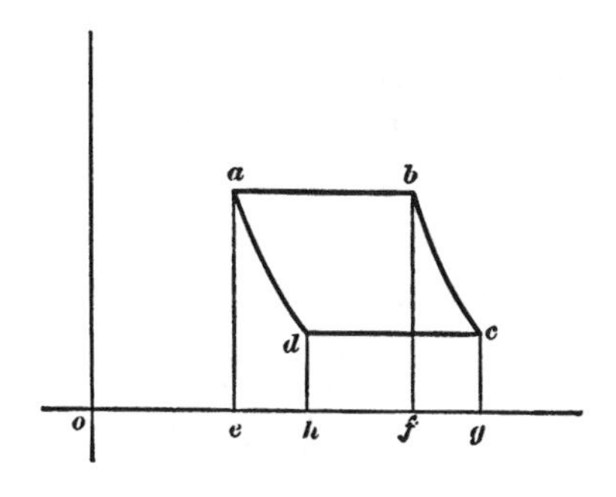

Fig. 3

condensation of the vapor will pass, so that the temperature remains constant and $= \tau$, in part without this body, so that the temperature rises. Let the operation be so managed that the first part of the compression is carried out only so far (to *oh*) that the volume *he* still remaining is exactly such that compression through it will raise the temperature from τ to t again. During the former diminution of volume the pressure remains invariable, $= gc$, and the external work employed is equal to the rectangle *gcdh*. During the latter diminution of volume the pressure increases and is represented by the curve *da*, which must end exactly at the point *a*, since the original pressure, *ea*, must correspond to the original temperature, t. The work employed in this last operation is $= hdae$. At the end of the operation the liquid and vapor are again in the same condition as at the beginning, so that the excess of the *external* work done over that employed is also the *total* work done. It is represented by the quadrilateral *abcd*, and its area must also be set equal to the *heat consumed* during the same time.

For our purposes we again assume that the changes just described are infinitely small, and on this assumption represent the whole process by the accompanying figure, in which the curves *ad* and *bc* which occur in Fig. 3 have become straight lines. So far as the area of the quadrilateral *abcd* is concerned, it may again be considered a parallelogram, and may be represented by the product $ef \cdot bk$.

If, now, the pressure of the vapor at the temperature t and at its maximum tension is represented by p, and if the temperature difference $t-\tau$ is represented by dt, we have

$$bk = \frac{dp}{dt}\,dt.$$

The line ef represents the increase of volume, which occurs in consequence of the passage of a certain quantity of liquid, which may be

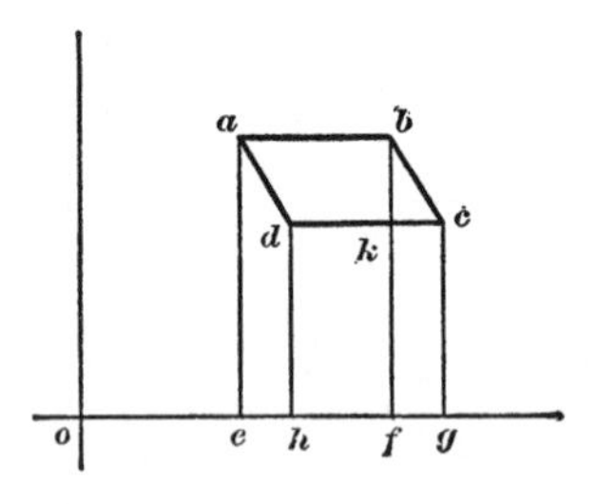

Fig. 4

called dm, over into vapor. Representing now the volume of a unit weight of the vapor at its maximum density at the temperature t by s, and the volume of the same quantity of liquid at the temperature t by σ, we have evidently

$$ef = (s-\sigma)\,dm,$$

and consequently the area of the quadrilateral, or

$$\text{(5)} \qquad \textit{The work done} = (s-\sigma)\frac{dp}{dt}\,dmdt.$$

In order to represent the quantities of heat concerned, we will introduce the following symbols. The quantity of heat which becomes latent when a unit weight of the liquid evaporates at the temperature t and under the corresponding pressure, is called r, and the specific heat of the liquid is called c. Both of these quantities, as well as also s, σ, and $\frac{dp}{dt}$, are to be considered functions of t. Finally, let us denote by hdt the quantity of heat which must be imparted to a unit weight of the vapor if its temperature is raised from t to $t+dt$, while it is so compressed that it is again at the maximum density for this temperature without the precipitation of any part of it. The quantity h is likewise a function of t. It will, for the present, be left undetermined whether it has a positive or negative value.

If we now denote by μ the mass of liquid originally present in the vessel, and by m the mass of vapor, and further by dm the mass which evaporates during the expansion from oe to of, and by $d'm$ the mass which condenses during the compression from og to oh, the heat which becomes latent in the first operation and is taken from the body A is

$$rdm,$$

and that which is set free in the second operation and is given to the body B is

$$\left(r - \frac{dr}{dt}\,dt\right) d'm.$$

In the other expansion and in the other compression heat is neither gained nor lost, so that, at the end of the process,

$$\text{(6)} \qquad \textit{The heat consumed} = rdm - \left(r - \frac{dr}{dt}\,dt\right) d'm.$$

In this expression the differential $d'm$ must be replaced by dm and dt. For this purpose we make use of the conditions under which the second expansion and the second compression occurred. The mass of vapor, which condenses during the compression from oh to oe, and which would be evolved by the corresponding expansion from oe to oh, may be represented by δm, and that which is evolved by the expansion from of to og by $\delta' m$. We then have at once, since at the end of the process the same mass of liquid μ and the mass of vapor m must be present as at the beginning, the equation

$$dm + \delta' m = d'm + \delta m.$$

Further, we obtain for the expansion from oe to oh, since in it the temperature of the mass of liquid μ and the mass of vapor m must be lowered by dt without the emission of heat, the equation

$$r\delta m - \mu \cdot cdt - m \cdot hdt = 0;$$

and similarly for the expansion from of to og, by substituting $\mu - dm$ and $m + dm$ for μ and m, and $\delta' m$ for δm,

$$r\delta' m - (\mu - dm)cdt - (m + dm)hdt = 0.$$

If from these three equations and (6) we eliminate the magnitudes $d'm$, δm, and $\delta' m$, and reject terms of higher order than the second, we have

$$\text{(7)} \qquad \textit{The heat consumed} = \left(\frac{dr}{dt}+c-h\right) dmdt.$$

The formulas (7) and (5) must now be connected in the same way as that used in the case of the permanent gases, that is,

$$\frac{\left(\frac{dr}{dt}+c-h\right) dmdt}{(s-\sigma)\frac{dp}{dt}dmdt} = A,$$

and we obtain as the analytical expression of the fundamental principle in the case of vapors at their maximum density the equation

$$\text{(III.)} \qquad \frac{dr}{dt}+c-h = A\,(s-\sigma)\,\frac{dp}{dt}.$$

If, instead of using our principle, we adopt the assumption that the quantity of heat is *constant*, we must replace (III.), as appears from (7), by

$$\text{(8)} \qquad \frac{dr}{dt}+c-h = 0.$$

This equation has been used, if not exactly in the same form, at least in its general sense, to obtain a value for the magnitude h. So long as Watt's law is considered true for water, that the sum of the free and latent heats of a quantity of vapor at its maximum density is equal for all temperatures, and that therefore

$$\frac{dr}{dt}+c = 0,$$

it must be concluded that for this liquid $h = 0$. This conclusion has, in fact, often been stated as correct, in that it has been said that if a quantity of vapor is at its maximum density, and then compressed or expanded in a vessel impermeable by heat, it remains at its maximum density. But since Regnault* has corrected Watt's law by substituting for it the approximate relation

$$\frac{dr}{dt}+c = 0.305,$$

* *Mém. de l'Acad.*, xxi, the 9th and 10th memoirs.

the equation (8) gives for h the value 0.305. It would therefore follow that the quantity of vapor formerly considered in the vessel impermeable by heat would be partly condensed by compression, and on expansion would not remain at the maximum density, since its temperature would not fall in a way to correspond to the diminution of pressure.

It is entirely different if we replace equation (8) by (III.). The expression on the right-hand side is, from its nature, always positive, and it therefore follows that h must be less than 0.305. It will subsequently appear that the value of this expression is so great that h is negative. We must therefore conclude that the quantity of vapor before mentioned is partly condensed, not by *compression*, but by *expansion*, and that by compression its temperature rises at a greater rate than the density increases, so that it does not remain at its maximum density.

It must be admitted that this result is exactly opposed to the common view already referred to; yet I do not believe that it is contradicted by any experimental fact. Indeed, it is more consistent than the former view with the behavior of steam as observed by Pambour. Pambour* found that the steam which issues from a locomotive after it has done its work always has the temperature at which the tension, observed at the same time, is a maximum. From this it follows either that $h = 0$, as it was once thought to be, because this assumption agreed with Watt's law, accepted as probably true, or that h is *negative*. For if h were *positive*, the temperature of the vapor, when released, would be too high in comparison with its tension, and that could not have escaped Pambour's notice. If, on the other hand, h is *negative*, according to our former statement, there can never arise from this cause too low a temperature, but a part of the steam must become liquid, so as to maintain the rest at the proper temperature. This part need not be great, since a small quantity of vapor sets free on condensation a relatively large quantity of heat, and the water formed will probably be carried on mechanically by the rest of the steam, and will in such researches pass unnoticed, the more likely as it might be thought, if it were to be observed, that it was water from the boiler carried out mechanically.

The results thus far obtained have been deduced from the fundamental principle without any further hypothesis. The equation

---

* *Traité des Locomotives*, second edition, and *Théorie des Machines à Vapeur*, second edition.

(II.a) obtained for permanent gases may, however, be made much more fruitful by the help of an obvious subsidiary hypothesis. The gases show in their various relations, especially in the relation expressed by the M. and G. law between volume, pressure, and temperature, so great a regularity of behavior that we are naturally led to take the view that the mutual attraction of the particles, which acts within solid and liquid bodies, no longer acts in gases, so that while in the case of other bodies the heat which produces expansion must overcome not only the external pressure but the internal attraction as well, in the case of gases it has to do only with the external pressure. If this is the case, then during the expansion of a gas only so much heat becomes *latent* as is used in doing *external* work. There is, further, no reason to think that a gas, if it expands at constant temperature, contains more *free* heat than before. If this be admitted, we have the law: *a permanent gas, when expanded at constant temperature, takes up only so much heat as is consumed in doing external work during the expansion.* This law is probably true for any gas with the same degree of exactness as that attained by the M. and G. law applied to it.

From this it follows at once that

$$\left(\frac{dQ}{dv}\right) = A \cdot R \frac{a+t}{v}, \tag{9}$$

since, as already noticed, $R \frac{a+t}{v} dv$ represents the external work done during the expansion dv. It follows that the function U which occurs in (II.a) does not contain v, and the equation therefore takes the form

$$dQ = cdt + AR \frac{a+t}{v} dv, \tag{II.b}$$

where c can be a function of t only. It is even probable that this magnitude c, which represents the specific heat of the gas at constant volume, is a constant.

Now in order to apply this equation to special cases, we must introduce the relation between the variables Q, t, and v, which is obtained from the conditions of each separate case, into the equation, and so make it integrable. We shall here consider only a few simple examples of this sort, which are either interesting in themselves or become so by comparison with other theorems already announced.

We may first obtain the specific heats of the gas at constant volume and at constant pressure if in (II.*b*) we set v=const., and p=const. In the former case, $dv=0$, and (II.*b*) becomes

$$\frac{dQ}{dt} = c. \tag{10}$$

In the latter case, we obtain from the condition p = const., by the help of equation (I.),

$$dv = \frac{Rdt}{p},$$

or

$$\frac{dv}{v} = \frac{dt}{a+t};$$

and this, substituted in (II.*b*), gives

$$\frac{dQ}{dt} = c' = c + AR, \tag{10a}$$

if we denote by c' the specific heat at constant pressure.

It appears, therefore, that the *difference of the two specific heats of any gas is a constant magnitude, AR.* This magnitude also involves a simple relation among the different gases. The complete expression for R is $\frac{p_0 v_0}{a+t_0}$, where p_0, v_0, and t_0 are any three corresponding values of p, v, and t for a unit of weight of the gas considered, and it therefore follows, as has already been mentioned in connection with the adoption of equation (I.), that R is inversely proportional to the specific gravity of the gas, and hence also that the same statement must hold for the difference $c'-c=AR$, since A is the same for all gases.

If we reckon the specific heat of the gas, not with respect to the unit of *weight*, but, as is more convenient, with respect to the unit of *volume*, we need only divide c and c' by v_0, if the volumes are taken at the temperature t_0 and pressure p_0. Designating these quotients by γ and γ', we obtain

$$\gamma' - \gamma = \frac{A \cdot R}{v_0} = A \frac{p_0}{a+t_0}. \tag{11}$$

In this last quantity nothing appears which is dependent on the particular nature of the gas, and *the difference of the specific heats referred to the unit of volume is therefore the same for all gases.*

This law was deduced by Clapeyron from Carnot's theory, though the constancy of the difference $c'-c$, which we have deduced before, is not found in this work, where the expression given for it still has the form of a function of the temperature.

If we divide equation (11) on both sides by γ, we have

$$k-1 = \frac{A}{\gamma}\cdot\frac{p_0}{a+t_0}, \tag{12}$$

in which k, for the sake of brevity, is used for the quotient $\frac{\gamma'}{\gamma}$, or, what amounts to the same thing, for the quotient $\frac{c'}{c}$. This quantity has acquired special importance in science from the theoretical discussion by Laplace of the propagation of sound in air. *The excess of this quotient over unity is therefore, for the different gases, inversely proportional to the specific heats of the same at constant volume, if these are referred to the unit of volume.* This law has, in fact, been found by Dulong from experiment* to be so nearly accurate that he has assumed it, in view of its theoretical probability, to be strictly accurate, and has therefore employed it, conversely, to calculate the specific heats of the different gases from the values of k determined by observation. It must, however, be remarked that the law is only theoretically justified when the M. and G. law holds, which is not the case with sufficient exactness for all the gases employed by Dulong.

If it is now assumed that the specific heat of gases at constant volume c is constant, which has been stated above to be very probable, the same follows for the specific heat at constant pressure, and consequently *the quotient of the two specific heats* $\frac{c'}{c}=k$ *is a constant.* This law, which Poisson has already assumed as correct on the strength of the experiments of Gay-Lussac and Welter, and has made the basis of his investigations on the tension and heat of gases,† is therefore in good agreement with our present theory, while it would not be possible on Carnot's theory as hitherto developed.

If in equation (II.b) we set $Q=$ const., we obtain the following equation between v and t:

$$cdt+A\cdot R\frac{a+t}{v}dv = 0, \tag{13}$$

---

* *Ann. de Chim. et de Phys.*, xli, and Pogg. *Ann.*, xvi.

† *Traité de Mécanique*, second edition, vol. ii, p. 646.

which gives, if c is considered constant,

$$v^{\frac{A \cdot R}{c}} \cdot (a+t) = \text{const.},$$

or, since from equation (10a), $\frac{AR}{c} = \frac{c'}{c} - 1 = k-1,$

$$v^{k-1}\,(a+t) = \text{const.}$$

Hence we have, if v_0, t_0, and p_0 are three corresponding values of v, t and p,

$$(14) \qquad \frac{a+t}{a+t_0} = \left(\frac{v_0}{v}\right)^{k-1}.$$

If we substitute in this relation the pressure p first for v and then for t by means of equation (I.), we obtain

$$(15) \qquad \left(\frac{a+t}{a+t_0}\right)^k = \left(\frac{p}{p_0}\right)^{k-1}$$

$$(16) \qquad \frac{p}{p_0} = \left(\frac{v_0}{v}\right)^k.$$

These are the relations which hold between volume, temperature, and pressure, if a quantity of gas is compressed or expanded within an envelope impermeable by heat. These equations agree precisely with those which have been developed by Poisson for the same case,* which depends upon the fact that he also treated k as a constant.

Finally, if we set $t=\text{const.}$ in equation (II.b), the first term on the right drops out, and there remains

$$(17) \qquad dQ = AR\frac{a+t}{v}\,dv,$$

from which we have

$$Q = AR(a+t)\log v + \text{const.},$$

or, if we denote by v_0, p_0, t_0, and Q_0 the values of v, p, t, and Q, which hold at the beginning of the change of volume,

$$(18) \qquad Q - Q_0 = AR(a+t_0)\log\frac{v}{v_0}.$$

From this follows the law also developed by Carnot: *If a gas changes its volume without changing its temperature, the quantities of heat evolved or*

---

* *Traité de Mécanique*, vol. ii, p. 647.

absorbed are in arithmetical progression, while the volumes are in geometrical progression.

Further, if we substitute for R in (18) the complete expression $\frac{p_0 v_0}{a+t_0}$, we have

$$Q - Q_0 = A p_0 v_0 \log \frac{v}{v_0}. \tag{19}$$

If now we apply this equation to the different gases, not by using equal *weights* of them, but such quantities as have at the outset equal volumes, v_0, it becomes in all its parts independent of the special nature of the gas, and agrees with the well-known law which Dulong proposed, guided by the above-mentioned simple relation of the magnitude $k-1$, *that all gases, if equal volumes of them are taken at the same temperature and under the same pressure, and if they are then compressed or expanded by an equal fraction of their volumes, either evolve or absorb an equal quantity of heat.* Equation (19) is, however, much more general. It states in addition, *that the quantity of heat is independent of the temperature at which the volume of the gas is altered*, if only the quantity of the gas employed is always determined so that the original volume v_0 is always the same at the different temperatures; and it states further, that *if the original pressure is different in the different cases, the quantities of heat are proportional to it.*

II. CONSEQUENCES OF CARNOT'S PRINCIPLE IN CONNECTION WITH THE ONE ALREADY INTRODUCED

Carnot assumed, as has already been mentioned, that *the equivalent of the work done by heat is found in the mere transfer of heat from a hotter to a colder body, while the quantity of heat remains undiminished.*

The latter part of this assumption—namely, that the quantity of heat remains undiminished—contradicts our former principle, and must therefore be rejected if we are to retain that principle. On the other hand, the first part may still obtain in all its essentials. For though we do not need a special equivalent for the work done, since we have assumed as such an actual *consumption* of heat, it still may well be possible that such a transfer of heat occurs *at the same time* as the consumption of heat, and also stands in a definite relation to the work done. It becomes important, therefore, to consider whether this assumption, besides the mere possibility, has also a sufficient probability in its favor.

A transfer of heat from a hotter to a colder body always occurs in those cases in which work is done by heat, and in which also the condition is fulfilled that the working substance is in the same state at the end as at the beginning of the operation. For example, we have seen, in the processes described above and represented in Figs. 1 and 3, that the gas and the evaporating water took up heat from the body A as their volume increased, and gave it up to the body B as their volume diminished; so that a certain quantity of heat was transferred from A to B, and this was in fact much greater than that which we assumed to be consumed, so that in the infinitely small changes, which are represented in Figs. 2 and 4, the latter was an infinitesimal of the second order, while the former was one of the first order. Yet, in order to establish a relation between the heat transferred and the work done, a certain restriction is necessary. For since a transfer of heat can take place without mechanical effect if a hotter and a colder body are immediately in contact and heat passes from one to the other by conduction, the way in which the transfer of a certain quantity of heat between two bodies at the temperatures t and τ can be made to do the maximum of work is so to carry out the process, as was done in the above cases, that two bodies of different temperatures never come in contact.

It is this *maximum* of work which must be compared with the heat transferred. When this is done it appears that there is in fact ground for asserting, with Carnot, that it depends only on the quantity of the heat transferred and on the temperatures t and τ of the two bodies A and B, but not on the nature of the substance by means of which the work is done. This maximum has, namely, the property that by *expending* it as great a quantity of heat can be transferred from the cold body B to the hot body A as passes from A to B when it is *produced*. This may easily be seen, if we think of the whole process formerly described as carried out in the reverse order, so that, for example, in the first case the gas first expands by itself, until its temperature falls from t to τ, is then expanded in contact with B, is then compressed by itself until its temperature is again t, and finally is compressed in contact with A. In this case more work will be employed during the compression than is produced during the expansion, so that on the whole there is a loss of work, which is exactly as great as the gain of work in the former process. Further, there will be just as much heat taken from the body B as was before given to it, and just as much given to the body A as was before taken from it, whence it follows not only that the same amount of heat is produced as was formerly consumed, but also that the heat which in the former process was transferred from A to B now passes from B to A.

If we now suppose that there are two substances of which the one can produce more work than the other by the transfer of a given amount of heat, or, what comes to the same thing, needs to transfer less heat from A to B to produce a given quantity of work, we may use these two substances alternately by producing work with one of them in the above process, and by expending work upon the other in the reverse process. At the end of the operations both bodies are in their original condition; further, the work produced will have exactly counterbalanced the work done, and therefore, by our former principle, the quantity of heat can have neither increased nor diminished. The only change will occur in the *distribution* of the heat, since more heat will be transferred from B to A than from A to B, and so on the whole heat will be transferred from B to A. By repeating these two processes alternately it would be possible, without any expenditure of force or any other change, to transfer as much heat as we please from a *cold* to a *hot* body, and this is not in accord with the other relations of heat, since it always shows a tendency to equalize temperature differences and therefore to pass from *hotter* to *colder* bodies.

It seems, therefore, to be *theoretically* admissible to retain the first and the really essential part of Carnot's assumptions, and to apply it as a second principle in conjunction with the first; and the correctness of this method is, as we shall soon see, established already in many cases by its *consequences*.

On this assumption we may express the maximum of work which can be produced by the transfer of a unit of heat from the body A at the temperature t to the body B at the temperature τ, as a function of t and τ. The value of this function must naturally be smaller as the difference $t-\tau$ is smaller, and when this is infinitely small ($=dt$) it must go over into the product of dt and a function of t only. For this latter case, with which we will concern ourselves for the present, the work may be expressed by the form $\frac{1}{C}\cdot dt$, where C is a function of t only.

In order to apply this result to the permanent gases, we return to the process represented in Fig. 2. In that case the quantity of heat,

$$\left(\frac{dQ}{dv}\right) dv,$$

passed during the first expansion from A to the gas, and by the first compression the part of it expressed by

$$\left[\left(\frac{dQ}{dv}\right)+\frac{d}{dv}\left(\frac{dQ}{dv}\right)\delta v-\frac{d}{dt}\left(\frac{dQ}{dv}\right)dt\right]d'v,$$

or by

$$\left(\frac{dQ}{dv}\right)dv-\left[\frac{d}{dt}\left(\frac{dQ}{dv}\right)-\frac{d}{dv}\left(\frac{dQ}{dt}\right)\right]dvdt,$$

was given up to the body B. The latter magnitude is, therefore, the quantity of heat transferred. Since we may neglect the term of the second order with respect to the one of the first order, we retain simply

$$\left(\frac{dQ}{dv}\right)dv.$$

The work produced at the same time was

$$\frac{Rdv\cdot dt}{v},$$

and we can thus at once form the equation

$$\frac{\frac{Rdv\cdot dt}{v}}{\left(\frac{dQ}{dv}\right)dv}=\frac{1}{C}\cdot dt,$$

or,

$$\text{(IV.)}\qquad \left(\frac{dQ}{dv}\right)=\frac{RC}{v}.$$

If, in the second place, we make a similar application to the process represented in Fig. 4 relating to vaporization, we have for the quantity of heat carried from A to B

$$(r-\frac{dr}{dt}dt)d'm,$$

or

$$rdm-\left(\frac{dr}{dt}+c-h\right)dmdt,$$

for which, by neglecting the term of the second order, we may set simply

$$rdm.$$

The work produced was

$$(s-\sigma)\,\frac{dp}{dt}\,dmdt,$$

and we therefore get the equation

$$\frac{(s-\sigma)\,\frac{dp}{dt}\cdot dm\cdot dt,}{rdm}=\frac{1}{C}\cdot dt,$$

or,

$$\text{(V.)}\qquad r=C\cdot(s-\sigma)\,\frac{dp}{dt}.$$

These are the two analytical expressions of Carnot's principle, as they are given by Clapeyron in his memoir, in a somewhat different form. For vapors he stops with this equation (V.) and some immediate applications of it. For gases, on the other hand, he makes the equation (IV.) the basis of a more extended development. It is in this development that the partial disagreement appears between his results and ours.

We shall now connect these two equations with the results of the first principle, first considering equation (IV.) in connection with the consequences formerly deduced for the case of permanent gases.

If we restrict ourselves to that result which depends only on the fundamental principle—that is, to equation (II.*a*)—we can use equation (IV.) to further define the magnitude U, which appears there as an arbitrary function of v and t, and our equation becomes

$$\text{(II.}c\text{)}\qquad dQ=\left[B+R\left(\frac{dC}{dt}-A\right)\log v\right]dt+\frac{R\cdot C}{v}\,dv,$$

where B is now an arbitrary function of t only.

If we also accept as correct the subsidiary hypothesis, then equation (IV.) is not necessary for the further definition of (II.*a*), since the same end is more completely attained by equation (9), which followed as an immediate consequence of this hypothesis in connection with the first principle. We gain, however, an opportunity to subject the results of the two principles to a comparative test. Equation (9) reads:

$$\left(\frac{dQ}{dv}\right)=\frac{R\cdot A(a+t)}{v},$$

and if we compare this with (IV.), we see that they both express the same result, only the one in a more definite way than the other, since for the general temperature function denoted in (IV.) by C, the equation (9) gives the special expression $A\ (a+t)$.

To this striking agreement it may be added that equation (V.), in which also the function C appears, confirms the view that $A\ (a+t)$ is the correct expression for this function. This equation has been used by Clapeyron and Thomson to calculate the values of C for several temperatures. Clapeyron chose as the temperatures the boiling points of ether, alcohol, water, and oil of turpentine, and by substituting in equation (V.) the values of $\frac{dp}{dt}$, s, and r for these liquids, determined by experiments at these boiling points, he obtained for C the numbers contained in the second column of the table which follows. Thomson, on the other hand, considered *water vapor* only, but at different temperatures, and thence calculated the value of C for every degree between 0° and 230° Cent. For this purpose Regnault's series of observations have given him an admissible basis so far as the magnitudes $\frac{dp}{dt}$ and r are concerned; but the magnitude s is not so well known for other temperatures as for the boiling point, and about this magnitude Thomson felt himself compelled to make an assumption, which he himself recognized as only approximately correct, and considered as a temporary aid, to be employed until more exact data are determined—namely, that water vapor at its maximum density follows the M. and G. law. The numbers which follow from his calculation for the same temperatures as those used by Clapeyron are given in the third column reduced to French units:

I

1 t in Cent. Degrees	2 C According to Clapeyron	3 C According to Thomson
35°.5	0.733	0.728
78°.8	0.828	0.814
100°	0.897	0.855
156°.8	0.930	0.952

It appears that the values of C found in both cases increase slowly with the temperature, similarly to the values of $A\ (a+t)$. They are in the ratio of the numbers in the following rows:

$$1 : 1.13 : 1.22 : 1.27$$
$$1 : 1.12 : 1.17 : 1.31$$

and if we determine the ratios of the values of $A\ (a+t)$ corresponding to the same temperatures, we obtain

$$1 : 1.14 : 1.21 : 1.39.$$

This series of *relative* values diverges from the two others only so far as can be accounted for by the uncertainty of the data which underlie them. The same agreement will be shown later in connection with the determination of the constant A, in respect to the *absolute* values.

Such an agreement between results which are obtained from entirely different principles cannot be accidental; it rather serves as a powerful confirmation of the two principles and the first subsidiary hypothesis annexed to them.

Returning again to the application of equations (IV.) and (V.), we may remark that the former, so far as relates to the permanent gases, has only served to confirm conclusions already obtained. In the consideration of vapors, and of all other substances to which Carnot's principle will be applied in the future, it furnishes, however, an essential improvement, in that it permits us to replace the function C, which recurs everywhere, by the definite expression $A\ (a+t)$.

By this substitution equation (V.) becomes

$$r = A\ (a+t) \cdot (s-\sigma) \frac{dp}{dt}, \tag{V.a}$$

and we therefore obtain for a vapor a simple relation between the temperature at which it is formed, the pressure, the volume, and the latent heat. This we can use in drawing further conclusions.

If the M. and G. law were accurate for vapors at their maximum density, we should have

$$ps = R(a+t). \tag{20}$$

Eliminating the magnitude s from (V.a) by the use of this equation, and neglecting the magnitude σ, which vanishes in comparison with s if the temperature is not very high, we obtain

$$\frac{1}{p}\frac{dp}{dt} = \frac{r}{AR\ (a+t)^2}.$$

If we make the further assumption that r is constant, we obtain by integration, if p_1 denotes the tension of the vapor at 100°,

$$\log \frac{p_0}{p_1} = \frac{r\,(t-100)}{A\cdot R(a+100)(a+t)},$$

or if we set $t-100 = \tau$, $a+100 = \alpha$, and $\frac{r}{A\cdot R(a+100)} = \beta$,

$$\log \frac{p_0}{p_1} = \frac{\beta\cdot\tau}{\alpha+\tau}. \tag{21}$$

This equation cannot, of course, be accurate, since the two assumptions made in its development are not accurate; but since these, at least to a certain extent, approach the truth, the quantity $\frac{\beta\cdot\tau}{\alpha+\tau}$ will roughly represent the value of the quantity $\log \frac{p_0}{p_1}$. We may explain in this way how it happens that this relation, if the constants α and β, instead of having values given them depending on their definitions, are considered as arbitrary, may serve as an empirical formula for the calculation of vapor tensions, without our being compelled to consider it as *fully* proved by theory, as is sometimes done.

The most immediate application of equation (V.*a*) is to *water vapor*, for which we have the largest collection of experimental data, in order to investigate *how far it departs, when at its maximum density, from the M. and G. law.* The magnitude of this departure cannot be unimportant, since carbonic acid and sulphurous acid, even at temperatures and tensions at which they are still far removed from their condensation points, show noticeable departures.

Equation (V.) may be put in the following form:

$$Ap\,(s-\sigma)\frac{a}{a+t} = \frac{ar}{(a+t)^2\,\frac{1}{p}\frac{dp}{dt}}. \tag{22}$$

The expression here found on the left-hand side would be very nearly constant, if the M. and G. law were applicable, since this law would give immediately, from (20),

$$A\cdot ps\,\frac{a}{a+t} = A\cdot Ra,$$

and $s-\sigma$ can be substituted for s in this equation with approximate accuracy. This expression is, therefore, especially suited to show clearly any departure from the M. and G. law, from the examination of its true values as they may be calculated from the expression on the right-hand side of (22). I have carried out this calculation for a series of temperatures, using for r and p the numbers given by Regnault.*

First with respect to the *latent heat*: Regnault states† that the quantity of heat λ, which must be imparted to a unit of weight of water, in order to heat it from 0° to t° and then to evaporate it at that temperature, may be represented with tolerable accuracy by the formula:

$$\lambda = 606.5 + 0.305\, t. \tag{23}$$

But now, from the significance of λ,

$$\lambda = r + \int_0^t c\,dt, \tag{23a}$$

and for the magnitude c, the specific heat of water, which appears in this formula, Regnault has given the formula:‡

$$c = 1 + 0.00004 \cdot t + 0.0000009 \cdot t^2. \tag{23b}$$

By the help of these two equations we obtain for the latent heat from equation (23) the expression:

$$r = 606.5 - 0.695 \cdot t - 0.00002 \cdot t^2 - 0.0000003 \cdot t^3. \S \tag{24}$$

Second, with respect to the pressure: in order to obtain from his numerous observations the most probable values, Regnault‖ made use of a graphic representation, by constructing curves, of which the

* *Mém. de l'Acad. de l'Inst. de France*, vol. xxi (1847).

† *Ibid.*, *Mém.* ix; also Pogg. *Ann.*, Bd. 98.

‡ *Ibid.*, *Mém.* x.

§ In most of his investigations Regnault has not so much observed the heat which becomes *latent* by evaporation of the vapor as that which becomes *free* by its condensation, and, therefore, since it has been shown above that, if the principle of the equivalence of heat and work is correct, the quantity of heat which a quantity of vapor gives up on condensation need not always be the same as that which it absorbs during its formation, the question may arise, whether such differences may not have entered in Regnault's experiments, so that the formula given for r would become inadmissible. I believe that we may answer this question in the negative, since Regnault so arranged his experiments that the condensation of the vapor occurred under the same pressure as its formation—that is, nearly under the pressure which corresponded as a maximum to the observed temperature, and in this case just as much heat must be evolved by condensation as is absorbed by evaporation.

‖ *Ibid.*, *Mém.* viii.

abscissae represented the temperature and the ordinates the pressure p, and which are drawn in sections from $-33°$ to $+230°$. From $100°$ to $230°$ he has also drawn a curve, of which the ordinates represent not p itself, by the logarithms of p. From this presentation the following values have been taken, which are to be considered as the immediate results of his observations, while the other *more complete* tables contained in the memoir were calculated from formulas, of which the choice and determination depended in the first instance upon these values:

II

t in Degrees Centigrade on the Air Thermometer	p in Milli-meters	t in Degrees Centigrade on the Air Thermometer	p in Millimeters	
			From the Curve of Numbers	From the Curve of Logarithms*
−20°	0.91	110°	1073.7	1073.3
−10	2.08	120	1489.0	1490.7
0	4.60	130	2029.0	2030.5
10	9.16	140	2713.0	2711.5
20	17.39	150	3572.0	3578.5
30	31.55	160	4647.0	4651.6
40	54.91	170	5960.0	5956.7
50	91.98	180	7545.0	7537.0
60	148.79	190	9428.0	9425.4
70	233.09	200	11660.0	11679.0
80	354.64	210	14308.0	14325.0
90	525.45	220	17390.0	17390.0
100	760.00	230	20915.0	20927.0

* Instead of the *logarithms* obtained immediately from the curve and adopted by Regnault, the *numbers* corresponding to them are given, in order to facilitate comparison with the numbers in the next column.

Now in order to carry out with these data the calculation in hand, I first determined from these tables the values of $\frac{1}{p}\cdot\frac{dp}{dt}$ for the temperatures $-15°$, $-5°$, $5°$, $15°$, etc., in the following way. Since the magnitude $\frac{1}{p}\cdot\frac{dp}{dt}$ only diminishes slowly as the temperature rises, I have considered as uniform the diminution in each interval of $10°$, say from $-20°$ to $-10°$, from $-10°$ to $0°$, etc., so that I could look on the value holding, for example, for $25°$ as the mean of all the values holding between 20° and 30°. On this assumption, since $\frac{1}{p}\cdot\frac{dp}{dt} = \frac{d(\log p)}{dt}$, I could use the formula:

$$\left(\frac{1}{p}\cdot\frac{dp}{dt}\right)_{25°} = \frac{\log p_{30°} - \log p_{20°}}{10},$$

or

$$\left(\frac{1}{p}\cdot\frac{dp}{dt}\right)_{25°} = \frac{\text{Log}\, p_{30°} - \text{Log}\, p_{20°}}{10 . M}, \tag{25}$$

in which Log indicates the Briggsian logarithms and M the modulus of this system. By help of these values of $\frac{1}{p}\cdot\frac{dp}{dt}$ and the values of r given by equation (24), and of the value 273 for a, the values which the expression on the right-hand side of (22), and so also the expression $Ap\,(s-\sigma)\,\frac{a}{a+t}$, take for the temperatures $-15°$, $-5°$, 5°, etc., were calculated and are given in the accompanying table. For temperatures above 100° both series of numbers given for p are used separately, and the two results found in each case given opposite each other. The significance of the third and fourth columns will be indicated in the sequel.

It appears at once from this table that $Ap\,(s-\sigma)\,\frac{a}{a+t}$ is not constant as it should be if the M. and G. law were applicable, but diminishes distinctly as the temperature rises. Between 35° and 90° this diminution appears to be very uniform. Below 35°, especially in the region of 0°, there appear noticeable irregularities, which, however, may be simply explained from the fact that in that region the pressure p and its differential coefficient $\frac{dp}{dt}$ are very small, and therefore small errors, which fall quite within the limits of the errors of observation, may become *relatively* important. It may be added that the curve by which the separate values of p are determined, as mentioned above, is not drawn in one stroke from $-35°$ to 100°, but, to economize space, is broken at 0°, so that at this temperature the progress of the curve cannot be determined so satisfactorily as it can within the separate portions below 0° and above 0°. From the way in which the differences occur in the foregoing table, it would seem that the value 4.60 mm taken for p at 0° is a little too great, since if that were so the values of $Ap\,(s-\sigma)\,\frac{a}{a+t}$ for

III

1 t in Degrees Centigrade on the Air Thermometer	$Ap(s-\sigma)\dfrac{a}{a+t}$ 2 From the Observed Values	 3 From Equation (27)	4 Differences
−15	30.61	30.61	0.00
−5	29.21	30.54	+1.33
5	30.93	30.46	−0.47
15	30.60	30.38	−0.22
25	30.40	30.30	−0.10
35	30.23	30.20	−0.03
45	30.10	30.10	0.00
55	29.98	30.00	+0.02
65	29.88	29.88	0.00
75	29.76	29.76	0.00
85	29.65	29.63	−0.02
95	29.49	29.48	−0.01
105	29.47 29.50	29.33	−0.14−0.17
115	29.16 29.02	29.17	+0.01+0.15
125	28.89 28.93	28.99	+0.10+0.06
135	28.88 29.01	28.80	−0.08−0.21
145	28.65 28.40	28.60	−0.05+0.20
155	28.16 28.25	28.38	+0.22+0.13
165	28.02 28.19	28.14	+0.12−0.05
175	27.84 27.90	27.89	+0.05−0.01
185	27.76 27.67	27.62	−0.14−0.05
195	27.45 27.20	27.33	−0.12+0.13
205	26.89 26.94	27.02	+0.13+0.08
215	26.56 26.79	26.68	+0.12−0.11
225	26.64 26.50	26.32	−0.32−0.18

the temperatures just under 0° would come out too small, and for those just over 0° too large. Above 100° the values of this expression do not diminish so regularly as between 35° and 95°; and yet they show, at least *in general*, a corresponding progress; and especially if we use a graphic representation, we find that the curve, which within that interval almost exactly joins the successive points determined by the numbers contained in the table, may be produced beyond that interval even to 230° quite naturally, so that these points are evenly distributed on both sides of it.

Within the range of the table the progress of the curve can be represented with fair accuracy by an equation of the form

$$Ap\,(s-\sigma)\frac{a}{a+t} = m - ne^{kt}, \tag{26}$$

where e is the base of the natural logarithms, and m, n, and k are constants. If these constants are calculated from the values which the curve gives for 45°, 125°, and 205°, we obtain:

$$(26a) \qquad m = 31.549, \quad n = 1.0486, \quad k = 0.007138,$$

and if for convenience we introduce the Briggsian logarithms, we obtain

$$(27) \quad \text{Log}\left[31.549 - Ap\,(s-\sigma)\frac{a}{a+t}\right] = 0.0206 + 0.003100\,t.$$

The numbers contained in the third column are calculated from this equation, and in the fourth are given the differences between these numbers and those in the second column.

From the foregoing we may easily deduce a formula by which we can more definitely determine the way in which the behavior of a vapor departs from the M. and G. law. By assuming this law to hold, and denoting by ps_0 the value of ps at 0°, we would have from (20),

$$\frac{ps}{ps_0} = \frac{a+t}{a},$$

and would have, therefore, for the differential coefficient $\frac{d}{dt}\left(\frac{ps}{ps_0}\right)$ a constant quantity—namely, the well-known coefficient of expansion $\frac{1}{a} = 0.003665$. Instead of this we have from (26), if we simply replace $s-\sigma$ by s, the equation:

$$(28) \qquad \frac{ps}{ps_0} = \frac{m - ne^{kt}}{m-n} \cdot \frac{a+t}{a},$$

and hence follows:

$$(29) \qquad \frac{d}{dt}\left(\frac{ps}{ps_0}\right) = \frac{1}{a} \cdot \frac{m - n\,[1 + k\,(a+t)]\,e^{kt}}{m-n}.$$

The differential coefficient is, therefore, not a constant, but a function of the temperature which diminishes as the temperature increases. If we substitute the numerical values of m, n, and k, given in (26a), we obtain, among others, the following values for this function:

IV

t	$\frac{d}{dt}\left(\frac{ps}{ps_0}\right)$	t	$\frac{d}{dt}\left(\frac{ps}{ps_0}\right)$	t	$\frac{d}{dt}\left(\frac{ps}{ps_0}\right)$
Deg.		*Deg.*		*Deg.*	
0	0.00342	70	0.00307	140	0.00244
10	0.00338	80	0.00300	150	0.00231
20	0.00334	90	0.00293	160	0.00217
30	0.00329	100	0.00285	170	0.00203
40	0.00325	110	0,00276	180	0.00187
50	0.00319	120	0.00266	190	0.00168
60	0.00314	130	0.00256	200	0.00149

It appears from this table that at low temperatures the departures from the M. and G. law are only slight, but that at higher temperatures—for example, at 100°, and upwards—they can no longer be neglected.

It may appear at first sight remarkable that the values found for $\frac{d}{dt}\left(\frac{ps}{ps_0}\right)$ are *smaller* than 0.003665, since we know that in the case of gases, especially of those, like carbonic acid and sulphurous acid, which deviate most widely from the M. and G. law, the coefficient of expansion is not *smaller*, but *greater*, than that number. We are not, however, justified in making an immediate comparison between the differential coefficients which we have just determined and the coefficient of expansion in the ordinary sense of the words, which relate to the increase of volume *at constant pressure*, nor yet with the number obtained by keeping *the volume constant* during the heating process, and then observing the increase in the expansive force. We are dealing here with a third special case of the general differential coefficient $\frac{d}{dt}\left(\frac{ps}{ps_0}\right)$—namely, with that which arises when, as the heating goes on, the pressure increases in the same proportion as it does with water vapor when it is kept at its maximum density; and we must consider carbonic acid in these relations if we wish to institute a comparison.

Water vapor has a tension of 1^m at about 108°, and of 2^m at $129\frac{1}{2}$°. We will, therefore, examine the behavior of carbonic acid if it is heated by $21\frac{1}{2}$°, and if the pressure upon it is at the same time increased from 1^m to 2^m. According to Regnault* the coefficient

* *Mém. de l'Acad., Mém.* i.

of expansion of carbonic acid at the constant pressure 760^{mm} is 0.003710, and at the pressure 2520^{mm} is 0.003846. For a pressure of 1500^{mm} (the mean between 1^{m} and 2^{m}), if we consider the increase of the coefficient of expansion as proportional to the increase of pressure, we obtain the value 0.003767. If carbonic acid were heated at this mean pressure from 0° to $21\frac{1}{2}$°, the magnitude $\frac{pv}{pv_0}$ would increase from 1 to $1+0.003767\times 21.5=1.08099$. Now from others of Regnault's researches* it is known that if carbonic acid, taken at a temperature near 0° under the pressure 1^{m}, is subjected to the pressure 1.98292^{m}, the magnitude ps decreases in the ratio of $1:0.99146$; so that for an increase of pressure from 1^{m} to 2^{m} there would be a decrease of this magnitude in the ratio of $1:0.99131$. If, now, both operations were performed at once—that is, the elevation of temperature from 0° to $21\frac{1}{2}$° and the increase in pressure from 1^{m} to 2^{m}—the magnitude $\frac{pv}{pv_0}$ would increase nearly from 1 to $1.08099\times 0.99131=1.071596$, and hence we obtain for the mean value of the differential coefficient $\frac{d}{dt}\left(\frac{pv}{pv_0}\right)$:

$$\frac{0.071596}{21.5}=0.00333.$$

It appears, therefore, that in the case now under consideration, a value is obtained for carbonic acid which is less than 0.003665, and therefore a similar result for a vapor at *its maximum density* should not be considered at all improbable.

If, on the other hand, we were to determine the real coefficient of expansion of the vapor—that is, the number which expresses by how much a quantity of vapor expands if it is taken at a certain temperature at its maximum density, and then removed from the water and heated under constant pressure—we should certainly obtain a value which would be *greater*, and perhaps *considerably* greater, than 0.003665.

From equation (26) we easily obtain the *relative* volumes of a unit of weight of vapor at its maximum density for different temperatures, referred to the volume at some definite temperature. In order to calculate the *absolute* volumes from these with sufficient precision, we must know the value of the constant A with greater accuracy than is as yet the case.

* *Mém. de l'Acad.*, *Mém.* vi.

The question now arises whether any one volume can be assigned with sufficient accuracy to permit its use as a starting point in the calculation of the other absolute values from the relative values. Many investigations of the specific weight of water vapor have been carried out, the results of which, however, are not, in my opinion, conclusive for the case with which we are now dealing, in which the vapor is at its maximum density. The numbers which are ordinarily given, especially the one obtained by Gay-Lussac—0.6235—agree very well with the theoretical value obtained by assuming that 2 parts of hydrogen and 1 part of oxygen combine to form 2 parts of water vapor—that is, with the value

$$\frac{2 \times 0.06926 + 1.10563}{2} = 0.622.$$

These numbers, however, are obtained from observations which were not carried out at temperatures at which the resulting pressure was equal to the maximum expansive force, but at higher temperatures. In this condition the vapor might nearly conform to the M. and G. law, and the agreement with the theoretical value may thus be explained. To pass from this result to the condition of maximum density by the use of the M. and G. law would contradict our previous conclusions, since Table IV shows too large a departure from this law, at the temperatures at which the determination was made, to make such a use of the law possible. Those experiments in which the vapor was observed at its maximum density give for the most part larger numbers, and Regnault has concluded* that even at a temperature a little over 30°, in the case in which the vapor is developed in vacuum, a sufficient agreement with the theoretical value is reached only when the tension of the vapor amounts to no more than 0.8 of that which corresponds to the observed temperature as the maximum. A definite conclusion, however, cannot be drawn from this observation, since it is doubtful, as Regnault remarks, whether the departure is really due to too great a specific weight of the vapor formed, or whether a quantity of water remained condensed on the walls of the glass globe. Other experiments, which were so executed that the vapor did not form in vacuum but saturated a current of air, gave results which were tolerably free from any irregularities,† yet even these results, important as they

* *Ann. de Chim. et de Phys.*, III Sér., xv, p. 148.

† *Ibid.*, p. 158 ff.

are in other relations, do not enable us to form any definite conclusions as to the behavior of vapor in a vacuum.

In this state of uncertainty the following considerations may perhaps be of some service in filling the gap. Table IV shows that the vapor at its maximum density conforms more closely to the M. and G. law as the temperature is lower, and it may hence be concluded that the specific weight will approach the theoretical value more nearly at lower than at higher temperatures. If therefore, for example, we assume the value 0.622 as correct for 0° and then calculate the corresponding value d for higher temperatures by the help of the following equation deduced from (26),

$$d = 0.622 \frac{m-n}{m-ne^{kt}}, \tag{30}$$

we obtain much more probable values than if we were to adopt 0.622 as correct for all temperatures. The following table presents some of these values:

V

t	0°	50°	100°	150°	200°
d	0.622	0.631	0.645	0.666	0.698

Strictly speaking, we must go further than this. In Table III we see that the values of $Ap\,(s-\sigma)\,\frac{a}{a+t}$, as the temperature falls, approach a limiting value, which is not reached even for the lowest temperatures of the table, and it is only for this limiting value that we have a right to assume the applicability of the M. and G. law and so set the specific weight equal to 0.622. The question therefore arises what this limiting value is. If we could consider the formula (26) as applicable for temperatures below $-15°$, we would have only to take the value which it approaches asymptotically, $m=31.549$, and we could then replace equation (30) by the equation

$$d = 0.622 \frac{m}{m-ne^{kt}}. \tag{31}$$

From this equation we obtain for the specific weight at 0° the value 0.643 instead of 0.622, and the other numbers of the preceding table must be increased in the same ratio. We are, however, not justified in so extended an application of formula (26), since it is only obtained empirically from the values given in Table III, and of these, those which relate to the lowest temperatures are rather uncertain. We must therefore, for the present, treat the limiting value of $A\,(s-\sigma)\,\frac{a}{a+t}$ as unknown, and content ourselves with such an approximation as the numbers in the preceding tables warrant. We may, however, conclude that these numbers are too small rather than too great.

If we combine equation (V.*a*) with equation (III.) deduced from the first principle, we may eliminate $A\,(s-\sigma)$, and obtain:

$$(32) \qquad \frac{dr}{dt}+c-h = \frac{r}{a+t}.$$

By means of this equation we may determine the magnitude h, which has already been stated to be negative. If we set for c and r the expressions given in (23*b*) and (24), and for a the number 273, we obtain:

$$(33) \quad h = 0.305 - \frac{606.5-0.695\,t-0.00002\,t^2-0.0000003\,t^3}{273+t},$$

and hence obtain for h, among others, the values:

VI

t	0°	50°	100°	150°	200°
h	−1.916	−1.465	−1.133	−0.879	−0.676

In a way similar to that which we have followed in the case of water vapor, we might apply equation (V.*a*) to the vapors of other liquids also, and then compare the results obtained for these different liquids, as has been done with the numbers calculated by Clapeyron and contained in Table I. We shall not, however, go into these applications any further at present.

We must now endeavor to determine, at least approximately, the numerical value of the constant A, or, what is more useful, of the fraction $\frac{1}{A}$, that is, *the work equivalent of the unit of heat.*

For this purpose we can first use equation (10*a*) *for the permanent gases,* which amounts to the same thing as the method already employed by Mayer and Helmholtz. This equation is:

$$c' = c+AR,$$

and if we set for c the equivalent expression $\frac{c'}{k}$, we have:

$$\frac{1}{A} = \frac{k \cdot R}{(k-1)\, c'}. \tag{34}$$

The value commonly taken for c' for atmospheric air from the researches of Delaroche and Bérard is 0.267, and for k from the researches of Dulong is 1.421. Further, to determine $R = \frac{p_0 v_0}{a + t_0}$, we know that the pressure of one atmosphere (760^{mm}) on a square meter is 10,333 kilograms, and that the volume of one kilogram of atmospheric air under that pressure and at the temperature of the freezing point $= 0.7733$ cubic meters. Hence follows:

$$R = \frac{10333 \cdot 0.7733}{273} = 29.26,$$

and consequently

$$\frac{1}{A} = \frac{1.421 \cdot 29.26}{0.421 \cdot 0.267} = 370,$$

that is, by the expenditure of a unit of heat (that quantity of heat which will raise the temperature of 1 kilogram of water from 0° to 1°) 370 kilograms can be lifted to the height of 1^{m}. Little confidence can be placed in this number, on account of the uncertainty of the numbers 0.267 and 1.421. Holtzmann gives as the limits, between which he is in doubt, 343 and 429.

We may further use the equation (V.a) developed for *vapors*. If we wish to apply it to *water* vapor, we can use the determinations given in the former part of our work, whose result is expressed in equation (26). If we choose in this equation the temperature 100°, for example, and set for p the corresponding pressure of 1 atmosphere $= 10,333$ kilograms, we obtain:

$$\frac{1}{A} = 257\,(s - \sigma). \tag{35}$$

If we now use Gay-Lussac's value of the specific weight of water vapor, 0.6235, we obtain $s = 1.699$, and hence,

$$\frac{1}{A} = 437.$$

Similar values are given by the use of the numbers contained in Table I, which Clapeyron and Thomson have calculated for C

from equation (V.). For if we consider these as the values of $A\,(a+t)$ for the temperatures corresponding to them, we obtain for $\frac{1}{A}$ a set of values which lie between 416 and 462.

It has already been mentioned that the specific weight of water vapor given by Gay-Lussac is probably somewhat too small for the case where the vapor is at its maximum density. The same may be said of most of the specific weights which are ordinarily given for other vapors. We must therefore conclude that the values of $\frac{1}{A}$ calculated from them are for the most part a little too great. If we take for water vapor the number 0.645 given in Table V, from which $s=1.638$, we obtain

$$\frac{1}{A} = 421.$$

This value is also perhaps a little, but probably not much, too great. We may therefore conclude, since this result should be given the preference over that obtained from atmospheric air, that *the work equivalent of the unit of heat is the lifting of something over* 400 *kilograms to the height of* 1^{m}.

We may now compare with this theoretical result those which Joule obtained in very different ways by direct observation. Joule obtained from the heat produced by magneto-electricity,

$$\frac{1}{A} = 460;*$$

from the quantity of heat which atmospheric air absorbs during its expansion,

$$\frac{1}{A} = 438,\dagger$$

and as a mean of a large number of experiments, in which the heat produced by friction of water, of mercury, and of cast iron, was observed,

$$\frac{1}{A} = 425.\ddagger$$

* *Phil. Mag.*, xxiii, p. 441. The number, given in English units, is reduced to French units.

† *Ibid.*, xxvi, p. 381.

‡ *Ibid.*, xxxv, p. 534.

The agreement of these three numbers, in spite of the difficulty of the experiments, leaves really no further doubt of the correctness of the fundamental principle of the equivalence of heat and work, and their agreement with the number 421 confirms in a similar way the correctness of Carnot's principle, in the form which it takes when combined with the first principle.

A CATALOG OF SELECTED

DOVER BOOKS

IN SCIENCE AND MATHEMATICS

Mathematics

FUNCTIONAL ANALYSIS (Second Corrected Edition), George Bachman and Lawrence Narici. Excellent treatment of subject geared toward students with background in linear algebra, advanced calculus, physics and engineering. Text covers introduction to inner-product spaces, normed, metric spaces, and topological spaces; complete orthonormal sets, the Hahn-Banach Theorem and its consequences, and many other related subjects. 1966 ed. 544pp. 6⅛ x 9¼. 0-486-40251-7

ASYMPTOTIC EXPANSIONS OF INTEGRALS, Norman Bleistein & Richard A. Handelsman. Best introduction to important field with applications in a variety of scientific disciplines. New preface. Problems. Diagrams. Tables. Bibliography. Index. 448pp. 5⅜ x 8½. 0-486-65082-0

VECTOR AND TENSOR ANALYSIS WITH APPLICATIONS, A. I. Borisenko and I. E. Tarapov. Concise introduction. Worked-out problems, solutions, exercises. 257pp. 5⅝ x 8¼. 0-486-63833-2

AN INTRODUCTION TO ORDINARY DIFFERENTIAL EQUATIONS, Earl A. Coddington. A thorough and systematic first course in elementary differential equations for undergraduates in mathematics and science, with many exercises and problems (with answers). Index. 304pp. 5⅜ x 8½. 0-486-65942-9

FOURIER SERIES AND ORTHOGONAL FUNCTIONS, Harry F. Davis. An incisive text combining theory and practical example to introduce Fourier series, orthogonal functions and applications of the Fourier method to boundary-value problems. 570 exercises. Answers and notes. 416pp. 5⅜ x 8½. 0-486-65973-9

COMPUTABILITY AND UNSOLVABILITY, Martin Davis. Classic graduate-level introduction to theory of computability, usually referred to as theory of recurrent functions. New preface and appendix. 288pp. 5⅜ x 8½. 0-486-61471-9

ASYMPTOTIC METHODS IN ANALYSIS, N. G. de Bruijn. An inexpensive, comprehensive guide to asymptotic methods–the pioneering work that teaches by explaining worked examples in detail. Index. 224pp. 5⅜ x 8½ 0-486-64221-6

APPLIED COMPLEX VARIABLES, John W. Dettman. Step-by-step coverage of fundamentals of analytic function theory–plus lucid exposition of five important applications: Potential Theory; Ordinary Differential Equations; Fourier Transforms; Laplace Transforms; Asymptotic Expansions. 66 figures. Exercises at chapter ends. 512pp. 5⅜ x 8½. 0-486-64670-X

INTRODUCTION TO LINEAR ALGEBRA AND DIFFERENTIAL EQUATIONS, John W. Dettman. Excellent text covers complex numbers, determinants, orthonormal bases, Laplace transforms, much more. Exercises with solutions. Undergraduate level. 416pp. 5⅜ x 8½. 0-486-65191-6

RIEMANN'S ZETA FUNCTION, H. M. Edwards. Superb, high-level study of landmark 1859 publication entitled "On the Number of Primes Less Than a Given Magnitude" traces developments in mathematical theory that it inspired. xiv+315pp. 5⅜ x 8½. 0-486-41740-9

CALCULUS OF VARIATIONS WITH APPLICATIONS, George M. Ewing. Applications-oriented introduction to variational theory develops insight and promotes understanding of specialized books, research papers. Suitable for advanced undergraduate/graduate students as primary, supplementary text. 352pp. 5⅜ x 8½. 0-486-64856-7

COMPLEX VARIABLES, Francis J. Flanigan. Unusual approach, delaying complex algebra till harmonic functions have been analyzed from real variable viewpoint. Includes problems with answers. 364pp. 5⅜ x 8½. 0-486-61388-7

AN INTRODUCTION TO THE CALCULUS OF VARIATIONS, Charles Fox. Graduate-level text covers variations of an integral, isoperimetrical problems, least action, special relativity, approximations, more. References. 279pp. 5⅜ x 8½. 0-486-65499-0

COUNTEREXAMPLES IN ANALYSIS, Bernard R. Gelbaum and John M. H. Olmsted. These counterexamples deal mostly with the part of analysis known as "real variables." The first half covers the real number system, and the second half encompasses higher dimensions. 1962 edition. xxiv+198pp. 5⅜ x 8½. 0-486-42875-3

CATASTROPHE THEORY FOR SCIENTISTS AND ENGINEERS, Robert Gilmore. Advanced-level treatment describes mathematics of theory grounded in the work of Poincaré, R. Thom, other mathematicians. Also important applications to problems in mathematics, physics, chemistry and engineering. 1981 edition. References. 28 tables. 397 black-and-white illustrations. xvii + 666pp. 6⅛ x 9¼. 0-486-67539-4

INTRODUCTION TO DIFFERENCE EQUATIONS, Samuel Goldberg. Exceptionally clear exposition of important discipline with applications to sociology, psychology, economics. Many illustrative examples; over 250 problems. 260pp. 5⅜ x 8½. 0-486-65084-7

NUMERICAL METHODS FOR SCIENTISTS AND ENGINEERS, Richard Hamming. Classic text stresses frequency approach in coverage of algorithms, polynomial approximation, Fourier approximation, exponential approximation, other topics. Revised and enlarged 2nd edition. 721pp. 5⅜ x 8½. 0-486-65241-6

INTRODUCTION TO NUMERICAL ANALYSIS (2nd Edition), F. B. Hildebrand. Classic, fundamental treatment covers computation, approximation, interpolation, numerical differentiation and integration, other topics. 150 new problems. 669pp. 5⅜ x 8½. 0-486-65363-3

THREE PEARLS OF NUMBER THEORY, A. Y. Khinchin. Three compelling puzzles require proof of a basic law governing the world of numbers. Challenges concern van der Waerden's theorem, the Landau-Schnirelmann hypothesis and Mann's theorem, and a solution to Waring's problem. Solutions included. 64pp. 5⅜ x 8½. 0-486-40026-3

THE PHILOSOPHY OF MATHEMATICS: AN INTRODUCTORY ESSAY, Stephan Körner. Surveys the views of Plato, Aristotle, Leibniz & Kant concerning propositions and theories of applied and pure mathematics. Introduction. Two appendices. Index. 198pp. 5⅜ x 8½. 0-486-25048-2

CATALOG OF DOVER BOOKS

INTRODUCTORY REAL ANALYSIS, A.N. Kolmogorov, S. V. Fomin. Translated by Richard A. Silverman. Self-contained, evenly paced introduction to real and functional analysis. Some 350 problems. 403pp. 5⅜ x 8½. 0-486-61226-0

APPLIED ANALYSIS, Cornelius Lanczos. Classic work on analysis and design of finite processes for approximating solution of analytical problems. Algebraic equations, matrices, harmonic analysis, quadrature methods, much more. 559pp. 5⅜ x 8½. 0-486-65656-X

AN INTRODUCTION TO ALGEBRAIC STRUCTURES, Joseph Landin. Superb self-contained text covers "abstract algebra": sets and numbers, theory of groups, theory of rings, much more. Numerous well-chosen examples, exercises. 247pp. 5⅜ x 8½. 0-486-65940-2

QUALITATIVE THEORY OF DIFFERENTIAL EQUATIONS, V. V. Nemytskii and V.V. Stepanov. Classic graduate-level text by two prominent Soviet mathematicians covers classical differential equations as well as topological dynamics and ergodic theory. Bibliographies. 523pp. 5⅜ x 8½. 0-486-65954-2

THEORY OF MATRICES, Sam Perlis. Outstanding text covering rank, nonsingularity and inverses in connection with the development of canonical matrices under the relation of equivalence, and without the intervention of determinants. Includes exercises. 237pp. 5⅜ x 8½. 0-486-66810-X

INTRODUCTION TO ANALYSIS, Maxwell Rosenlicht. Unusually clear, accessible coverage of set theory, real number system, metric spaces, continuous functions, Riemann integration, multiple integrals, more. Wide range of problems. Undergraduate level. Bibliography. 254pp. 5⅜ x 8½. 0-486-65038-3

MODERN NONLINEAR EQUATIONS, Thomas L. Saaty. Emphasizes practical solution of problems; covers seven types of equations. ". . . a welcome contribution to the existing literature...."–*Math Reviews*. 490pp. 5⅜ x 8½. 0-486-64232-1

MATRICES AND LINEAR ALGEBRA, Hans Schneider and George Phillip Barker. Basic textbook covers theory of matrices and its applications to systems of linear equations and related topics such as determinants, eigenvalues and differential equations. Numerous exercises. 432pp. 5⅜ x 8½. 0-486-66014-1

LINEAR ALGEBRA, Georgi E. Shilov. Determinants, linear spaces, matrix algebras, similar topics. For advanced undergraduates, graduates. Silverman translation. 387pp. 5⅜ x 8½. 0-486-63518-X

ELEMENTS OF REAL ANALYSIS, David A. Sprecher. Classic text covers fundamental concepts, real number system, point sets, functions of a real variable, Fourier series, much more. Over 500 exercises. 352pp. 5⅜ x 8½. 0-486-65385-4

SET THEORY AND LOGIC, Robert R. Stoll. Lucid introduction to unified theory of mathematical concepts. Set theory and logic seen as tools for conceptual understanding of real number system. 496pp. 5⅜ x 8¼. 0-486-63829-4

Physics

OPTICAL RESONANCE AND TWO-LEVEL ATOMS, L. Allen and J. H. Eberly. Clear, comprehensive introduction to basic principles behind all quantum optical resonance phenomena. 53 illustrations. Preface. Index. 256pp. 5⅜ x 8½. 0-486-65533-4

QUANTUM THEORY, David Bohm. This advanced undergraduate-level text presents the quantum theory in terms of qualitative and imaginative concepts, followed by specific applications worked out in mathematical detail. Preface. Index. 655pp. 5⅜ x 8½. 0-486-65969-0

ATOMIC PHYSICS (8th EDITION), Max Born. Nobel laureate's lucid treatment of kinetic theory of gases, elementary particles, nuclear atom, wave-corpuscles, atomic structure and spectral lines, much more. Over 40 appendices, bibliography. 495pp. 5⅜ x 8½. 0-486-65984-4

A SOPHISTICATE'S PRIMER OF RELATIVITY, P. W. Bridgman. Geared toward readers already acquainted with special relativity, this book transcends the view of theory as a working tool to answer natural questions: What is a frame of reference? What is a "law of nature"? What is the role of the "observer"? Extensive treatment, written in terms accessible to those without a scientific background. 1983 ed. xlviii+172pp. 5⅜ x 8½. 0-486-42549-5

AN INTRODUCTION TO HAMILTONIAN OPTICS, H. A. Buchdahl. Detailed account of the Hamiltonian treatment of aberration theory in geometrical optics. Many classes of optical systems defined in terms of the symmetries they possess. Problems with detailed solutions. 1970 edition. xv + 360pp. 5⅜ x 8½. 0-486-67597-1

PRIMER OF QUANTUM MECHANICS, Marvin Chester. Introductory text examines the classical quantum bead on a track: its state and representations; operator eigenvalues; harmonic oscillator and bound bead in a symmetric force field; and bead in a spherical shell. Other topics include spin, matrices, and the structure of quantum mechanics; the simplest atom; indistinguishable particles; and stationary-state perturbation theory. 1992 ed. xiv+314pp. 6⅛ x 9¼. 0-486-42878-8

LECTURES ON QUANTUM MECHANICS, Paul A. M. Dirac. Four concise, brilliant lectures on mathematical methods in quantum mechanics from Nobel Prize-winning quantum pioneer build on idea of visualizing quantum theory through the use of classical mechanics. 96pp. 5⅜ x 8½. 0-486-41713-1

THIRTY YEARS THAT SHOOK PHYSICS: THE STORY OF QUANTUM THEORY, George Gamow. Lucid, accessible introduction to influential theory of energy and matter. Careful explanations of Dirac's anti-particles, Bohr's model of the atom, much more. 12 plates. Numerous drawings. 240pp. 5⅜ x 8½. 0-486-24895-X

ELECTRONIC STRUCTURE AND THE PROPERTIES OF SOLIDS: THE PHYSICS OF THE CHEMICAL BOND, Walter A. Harrison. Innovative text offers basic understanding of the electronic structure of covalent and ionic solids, simple metals, transition metals and their compounds. Problems. 1980 edition. 582pp. 6⅛ x 9¼. 0-486-66021-4

HYDRODYNAMIC AND HYDROMAGNETIC STABILITY, S. Chandrasekhar. Lucid examination of the Rayleigh-Benard problem; clear coverage of the theory of instabilities causing convection. 704pp. 5⅜ x 8¼. 0-486-64071-X

INVESTIGATIONS ON THE THEORY OF THE BROWNIAN MOVEMENT, Albert Einstein. Five papers (1905–8) investigating dynamics of Brownian motion and evolving elementary theory. Notes by R. Fürth. 122pp. 5⅜ x 8½. 0-486-60304-0

THE PHYSICS OF WAVES, William C. Elmore and Mark A. Heald. Unique overview of classical wave theory. Acoustics, optics, electromagnetic radiation, more. Ideal as classroom text or for self-study. Problems. 477pp. 5⅜ x 8½. 0-486-64926-1

GRAVITY, George Gamow. Distinguished physicist and teacher takes reader-friendly look at three scientists whose work unlocked many of the mysteries behind the laws of physics: Galileo, Newton, and Einstein. Most of the book focuses on Newton's ideas, with a concluding chapter on post-Einsteinian speculations concerning the relationship between gravity and other physical phenomena. 160pp. 5⅜ x 8½. 0-486-42563-0

PHYSICAL PRINCIPLES OF THE QUANTUM THEORY, Werner Heisenberg. Nobel Laureate discusses quantum theory, uncertainty, wave mechanics, work of Dirac, Schroedinger, Compton, Wilson, Einstein, etc. 184pp. 5⅜ x 8½. 0-486-60113-7

ATOMIC SPECTRA AND ATOMIC STRUCTURE, Gerhard Herzberg. One of best introductions; especially for specialist in other fields. Treatment is physical rather than mathematical. 80 illustrations. 257pp. 5⅜ x 8½. 0-486-60115-3

AN INTRODUCTION TO STATISTICAL THERMODYNAMICS, Terrell L. Hill. Excellent basic text offers wide-ranging coverage of quantum statistical mechanics, systems of interacting molecules, quantum statistics, more. 523pp. 5⅜ x 8½. 0-486-65242-4

THEORETICAL PHYSICS, Georg Joos, with Ira M. Freeman. Classic overview covers essential math, mechanics, electromagnetic theory, thermodynamics, quantum mechanics, nuclear physics, other topics. First paperback edition. xxiii + 885pp. 5⅜ x 8½. 0-486-65227-0

PROBLEMS AND SOLUTIONS IN QUANTUM CHEMISTRY AND PHYSICS, Charles S. Johnson, Jr. and Lee G. Pedersen. Unusually varied problems, detailed solutions in coverage of quantum mechanics, wave mechanics, angular momentum, molecular spectroscopy, more. 280 problems plus 139 supplementary exercises. 430pp. 6½ x 9¼. 0-486-65236-X

THEORETICAL SOLID STATE PHYSICS, Vol. 1: Perfect Lattices in Equilibrium; Vol. II: Non-Equilibrium and Disorder, William Jones and Norman H. March. Monumental reference work covers fundamental theory of equilibrium properties of perfect crystalline solids, non-equilibrium properties, defects and disordered systems. Appendices. Problems. Preface. Diagrams. Index. Bibliography. Total of 1,301pp. 5⅜ x 8½. Two volumes. Vol. I: 0-486-65015-4 Vol. II: 0-486-65016-2

WHAT IS RELATIVITY? L. D. Landau and G. B. Rumer. Written by a Nobel Prize physicist and his distinguished colleague, this compelling book explains the special theory of relativity to readers with no scientific background, using such familiar objects as trains, rulers, and clocks. 1960 ed. vi+72pp. 5⅜ x 8½. 0-486-42806-0

CATALOG OF DOVER BOOKS

A TREATISE ON ELECTRICITY AND MAGNETISM, James Clerk Maxwell. Important foundation work of modern physics. Brings to final form Maxwell's theory of electromagnetism and rigorously derives his general equations of field theory. 1,084pp. 5⅜ x 8½. Two-vol. set. Vol. I: 0-486-60636-8 Vol. II: 0-486-60637-6

QUANTUM MECHANICS: PRINCIPLES AND FORMALISM, Roy McWeeny. Graduate student-oriented volume develops subject as fundamental discipline, opening with review of origins of Schrödinger's equations and vector spaces. Focusing on main principles of quantum mechanics and their immediate consequences, it concludes with final generalizations covering alternative "languages" or representations. 1972 ed. 15 figures. xi+155pp. 5⅜ x 8½. 0-486-42829-X

INTRODUCTION TO QUANTUM MECHANICS With Applications to Chemistry, Linus Pauling & E. Bright Wilson, Jr. Classic undergraduate text by Nobel Prize winner applies quantum mechanics to chemical and physical problems. Numerous tables and figures enhance the text. Chapter bibliographies. Appendices. Index. 468pp. 5⅜ x 8½. 0-486-64871-0

METHODS OF THERMODYNAMICS, Howard Reiss. Outstanding text focuses on physical technique of thermodynamics, typical problem areas of understanding, and significance and use of thermodynamic potential. 1965 edition. 238pp. 5⅜ x 8½. 0-486-69445-3

THE ELECTROMAGNETIC FIELD, Albert Shadowitz. Comprehensive undergraduate text covers basics of electric and magnetic fields, builds up to electromagnetic theory. Also related topics, including relativity. Over 900 problems. 768pp. 5⅜ x 8¼. 0-486-65660-8

GREAT EXPERIMENTS IN PHYSICS: FIRSTHAND ACCOUNTS FROM GALILEO TO EINSTEIN, Morris H. Shamos (ed.). 25 crucial discoveries: Newton's laws of motion, Chadwick's study of the neutron, Hertz on electromagnetic waves, more. Original accounts clearly annotated. 370pp. 5⅜ x 8½. 0-486-25346-5

EINSTEIN'S LEGACY, Julian Schwinger. A Nobel Laureate relates fascinating story of Einstein and development of relativity theory in well-illustrated, nontechnical volume. Subjects include meaning of time, paradoxes of space travel, gravity and its effect on light, non-Euclidean geometry and curving of space-time, impact of radio astronomy and space-age discoveries, and more. 189 b/w illustrations. xiv+250pp. 8⅜ x 9¼. 0-486-41974-6

STATISTICAL PHYSICS, Gregory H. Wannier. Classic text combines thermodynamics, statistical mechanics and kinetic theory in one unified presentation of thermal physics. Problems with solutions. Bibliography. 532pp. 5⅜ x 8½. 0-486-65401-X

Paperbound unless otherwise indicated. Available at your book dealer, online at **www.doverpublications.com**, or by writing to Dept. GI, Dover Publications, Inc., 31 East 2nd Street, Mineola, NY 11501. For current price information or for free catalogues (please indicate field of interest), write to Dover Publications or log on to **www.doverpublications.com** and see every Dover book in print. Dover publishes more than 500 books each year on science, elementary and advanced mathematics, biology, music, art, literary history, social sciences, and other areas.